VILLAGE OF THE DAMMED

Village of the Dammed

THE FIGHT FOR OPEN SPACE AND THE FLOODING OF A CONNECTICUT TOWN

James Lomuscio

University Press of New England
Hanover and Lebanon

Published by University Press of New England
One Court Street, Lebanon, NH 03766
www.upne.com
© by James Lomusico
Printed in the United States of America
5 4 3 2 1

All rights reserved. No part of this book may be reproduced in any form or by any electronic or mechanical means, including storage and retrieval systems, without permission in writing from the publisher, except by a reviewer, who may quote brief passages in a review. Members of educational institutions and organizations wishing to photocopy any of the work for classroom use, or authors and publishers who would like to obtain permission for any of the material in the work, should contact Permissions, University Press of New England, One Court Street, Lebanon, NH 03766.

Library of Congress Cataloging-in-Publication Data
Lomuscio, James, 1955–
   The fight for open space and the flooding of a Connecticut town / James Lomuscio.—1st ed.
     p. cm.
Includes bibliographical references and index.
ISBN 1-58465-477-5 (cloth : alk. paper)
   1. Public lands—Connecticut. 2. Conservation of natural resources—Connecticut—Citizen participation. I. Title.
HD243.C8 L66 2005
333.91'62'097469—dc22   2005003213

For My Father

Fig. 1. "Saugatuck Reservoir at Low Tide." Painting by Mary Ann Barr.

We believe the lake that will flood the Saugatuck Valley will be far more beautiful than the long rows of small cottages spotted along the river up through the length of the valley. . . . I would also like to say just a word concluding about our real estate operations. We have never bought any land whatsoever for the purpose of speculation.
   —Samuel P. Senior, president of the Bridgeport
   Hydraulic Company, *Westporter-Herald*,
   October 29, 1937.

When Mr. Bergschneider came along with his proposal [for a housing and golf course development], we were excited.
   —Daniel Neaton, BHC's vice president of real estate,
   New York Times, August 31, 1997.

"The thing I don't want to regret is for my daughter and my grandson not to have this. We never really appreciate what we have until it is lost."
   —Paul Newman, June 2000 press conference.

 CONTENTS

Foreword By James Prosek xi

Acknowledgments xiii

Introduction 1

1. Piercing the Surface 5
2. A Personal Connection 11
3. A Valley Blooms 14
4. A Lens on the Past 20
5. A Water Company Rises 32
6. Equity Versus Inequity 40
7. The Fighters 47
8. Road Wars and Water Rights 55
9. The Defeat 63
10. The Dammed 68
11. At Peace at the Reservoir 81
12. Another Valley, Another Time 85
13. Casting About 89
14. Into the Woods, Again 100

15. Celebrity Status   107

16. Letters to the Editor   112

17. Artful Strategies   122

18. Friends in High Places   126

    Epilogue   135

    Works Cited and Consulted   141

    Index   145

 **FOREWORD**

By James Prosek

In this fascinating book, James Lomuscio chronicles the history of Connecticut's flooded and forgotten Valley Forge and the 1930s struggles between a handful of towns in Fairfield County—Weston, Westport, and Redding among them—and a private utility, the Bridgeport Hydraulic Company.

The water company's history has been inextricably linked to the area since 1857 when it was founded, to supply water for major fires to a drinking water company, which would develop to service a booming Long Island Sound county within close proximity to New York City. Bridgeport Hydraulic's early president, none other than Bridgeport mayor and circus creator P. T. Barnum, built a secure foundation for the company. Then came Samuel Senior, who eventually would prevail in a fight with the Weston community, to flood Valley Forge—once thought to be among the most beautiful valleys in New England—to create the Saugatuck Reservoir.

From the beginning, the water company's motives were in question. They paid fewer taxes to the town for their land than private residents did. When they did suceed in gaining approval to build the reservoir, Valley Forge residents said the Bridgeport Hydraulic Company (BHC) did not pay fair market value for the land (an old charter gave Bridgeport Hydraulic the right of eminent domain to any waterway in the area). It was all performed in the name of the public good, but decades later the company tried to sell that same land for a major profit. Some asked whether the BHC was in the business of water or real estate.

The multiple ironies made apparent by the passage of time are what make this book especially compelling, as James Lomuscio fast forwards us from Valley Forge to a successful fight in the late 1990s to save the adjacent Trout Brook Valley. Today, the land around and beneath the several reservoirs, taken from the people whose homes were flooded, is preserved again for the public good—thanks to the efforts of conservationists and state officials. And Bridgeport, the booming city that was to benefit most and was expected to be among the richest in New England, suffered and sputtered to a halt when its industry left it. A character like Samuel Senior, despised for pushing through plans for the damming of the Saugatuck River, using all his influence in the then powerful Bridgeport courts, seems an unexpected hero sixty-five years later. If the valley hadn't been flooded and protected for drinking water, might it not all be developed today?

We who enjoy open spaces have a wish to preserve the beautiful place we live in. As I grew up in Easton, the reservoirs, though off limits to the public, were always my private playground. As a boy I took for granted what was available to me beyond those "No Tresspassing" signs. Only later did I realize what a treasure this land was, so wild and so close to New York City. The BHC, whatever its intentions from the beginning, was perhaps the best thing that could have happened to the area.

After reading this book, which brings us to the present, you will be left wondering what role the state, conservation groups, and the water company will play in the future preservation of these unique watershed lands, ones that my father always referred to as an "oasis." One can only hope that all will take those next necessary steps toward keeping this area abundantly beautiful.

 ACKNOWLEDGMENTS

I would like to thank all of those who helped me in my research for this book. I would specifically like to acknowledge Mary Ann Barr of the Weston Historical Society for making the society's photo and clipping archives available to me, and Weston resident Gary Samuelson for sharing his one-of-a-kind Valley Forge photos he had discovered on photographic glass plates that had been slated for the dump. I am also indebted to Weston's Jim Daniel and the late Jim Hoe, both time-honored historians with a passion for research. And special thanks to Easton residents Bob Kranyik and Princie Falkenhagen for their input and photos.

# VILLAGE OF THE DAMMED

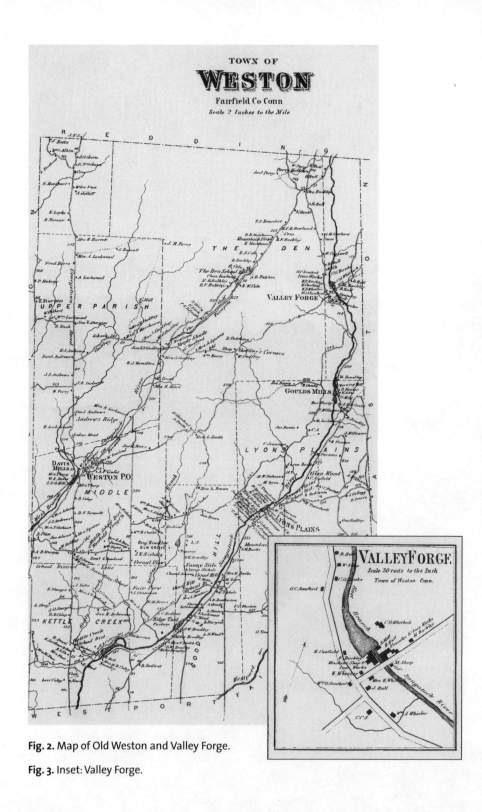

Fig. 2. Map of Old Weston and Valley Forge.

Fig. 3. Inset: Valley Forge.

Introduction

Miniature waves, the kind that glisten with specks of gold, break gently on mossy stones. From a damp boulder, I spin cast and hope for trout, bass, walleye, or whatever else is out there. My son casts, too. Behind us and under a tall hemlock, my father looks on, a walking stick in his hand, his cap defiant against the breeze.

Under a canopy of azure and as far as the eye can see, there's nothing but water and the green embrace of pines and hardwoods along the circuitous shoreline. And there's that small, undeveloped island off in the distance, one that's so intriguing and inviting but off limits.

Everything is quiet, blanketed with the serenity that belongs to nature alone.

"I feel like I'm somewhere in the Adirondacks," my father says.

"This is heaven," I say.

"It's better than heaven," my son chimes in with deliberate hyperbole.

The place is the 880-acre Saugatuck Reservoir in northern Weston and parts of Redding and Easton. And, yes, it could be somewhere in the Adirondacks. It resembles a lake born from springs of glacial deposits. The water is crystal clear. The finfish and the shells of freshwater clams seem primordial. The reservoir, surrounded by thousands of acres of water company watershed land, appears timeless.

Fig. 4a. and 4b. The Saugatuck Reservoir today. Courtesy Weston Historical Society, Inc., Herb Day Collection.

But actually, it's quite young, just sixty years old—much younger than my father who is so impressed by its ageless beauty.

Yes, these waters are as deceptive as they are deep in areas. And thus a story unfolds, a true tale about what lies beneath. It is the story of children, parents, and families who forged the American Dream here. It is the tale of homes, a school, a store, bustling factories and farms. And it's the story of how this community had its dreams dashed by the floodgates of then presumed progress.

The narrative begins with haunting images retrieved from photographic glass plates recovered from an old chicken coop-turned-garage. And it's best told when the sky is overcast and a mist creeps along the water's surface. A village known as Valley Forge flourished here between the nineteenth and early twentieth centuries, and its factories were on the cutting edge of iron and steel manufacturing. From hoes to oxcart wheels, myriad iron products were made here, horse drawn to Westport and shipped by sailing vessels to Southern states. And during the Civil War production peaked as the mills worked overtime.

But times change, and by the close of the 1800s, Weston's rise as an industrial center had ebbed. In the early 1900s, most of the iron-and-steel industry had moved to the Great Lakes. And by the late 1930s, Valley Forge's remaining residents, weary from the fight to save their homes from eminent domain, succumbed to the Bridgeport Hydraulic Company with its own plans for that scenic valley.

Some of the houses were moved. Others were dismantled, some even burned. Then in 1940 the river was dammed by a more than 110-foot wall of concrete, and this community vanished more than one hundred feet beneath what is now the Saugatuck Reservoir.

"It's all underwater, it's our lost Atlantis," says Jim Daniel, a former Weston first selectman and a Weston Historical Society member who chronicled the history of Valley Forge. "No kiss from Prince Charming or squad of lithe skin divers is going to bring back this sleeping beauty."

Yes, it's all been consigned to the memories of old-timers, who themselves are diminishing in number. To old photographs. To those who see our future in our past. To anyone curious about what lies beneath.

1

Piercing the Surface

To examine the Valley Forge saga is to look into the eyes of the Yankee soul, a spirit that wells up from below the placid surface of the Saugatuck Reservoir to the hillsides that surround it. It is to rethink each stone painstakingly laid in the mortarless walls that come to an abrupt halt at the water's edge. And it is to hear the still defiant roar of the Saugatuck River as it starts up again below the Samuel P. Senior Dam at the reservoir's southern-most point.

The story of Valley Forge, the reservoir and the water company lands that surround it, is more than just a ghostly tale of a submerged village. It is not just about a poor community losing its basic right to exist to an all powerful utility company, a David and Goliath story in which David loses. It is about the indomitable Yankee spirit, how it was left for dead—only to resurface with a vengeance more than a half-century later.

Valley Forge also raises familiar issues in Connecticut, ones that have come to the fore over the past decade as water companies, bolstered by filtration plants, have begun selling off watershed land no longer deemed necessary. In 1997, for example, BHC, then owned by Aquarion, flirted with the idea of selling a 730–acre watershed parcel known as Trout Brook Valley in Weston and Easton. The land would have gone to the developer National Fairways for a gated community

Fig. 5. Reservoir sign identifying the Samuel P. Senior Dam, named for the Bridgeport Hydraulic Company president who had it constructed. Courtesy Weston Historical Society, Inc.

of high-end housing and a private golf course with membership fees starting at $100,000.

But local opposition led by actor-philanthropist Paul Newman swelled all the way to the state level and garnered national interest. Along with a plethora of environmental concerns, since the land was near the Saugatuck Reservoir, came outrage. This land, which the water company had planned to sell for the benefit of its shareholders, had been seized years ago in the name of the public good around the same time Valley Forge met its watery fate.

"Bridgeport Hydraulic obtained this land either using eminent domain, which was granted to them by the legislature, or in the shadow of eminent domain many years ago, to put together a large parcel of land for a public purpose," Curtis Johnson, staff attorney for the New Haven–based Connecticut Fund for the Environment, told me in an interview for the *New York Times* in July 1998. He added that under the development plan for Trout Brook there would be little public benefit.

As I walked the Trout Brook acreage near the reservoir that summer, I came upon old foundations and worm-eaten house beams.

Though I hadn't even heard the story of Valley Forge yet, I couldn't help but feel sad. These had been people's homes.

Robert Kranyik, an Easton resident and one of the founders of the Coalition to Save Trout Brook, said the historical lessons of Valley Forge steeled his group's effort. Refusing to repeat the past, they remained steadfast. In 1999, with the help of Mr. Newman's celebrity status, the coalition scored a major environmental victory. Trout Brook Valley was purchased as open space. This win encouraged preservationists a year later when faced with a much greater threat. The British-based Kelda corporation, which had then just purchased and still owns Aquarion, had planned to sell much of its prime and pristine forest, totaling more than 18,000 acres in Fairfield and Litchfield counties, for housing development. Again, Mr. Newman, local officials and state legislators led the charge. And again, David hurled a well-aimed stone at Goliath as the state purchased the land. These latter-day victories prove the loss of Valley Forge was not in vain.

Weston, Connecticut, lies just forty-five miles from the New York City line. A metropolitan bedroom community of 10,037, it fiercely clings to its rural identity, with two-acre zoning, only a handful of stores in its town center, no sidewalks, no sewers, and more than 2,400 acres of open space. A old onion barn near the town center is a revered relic of Weston's agrarian past and the place where banners and posters of town events are still nailed weekly to its broad side.

But vestiges of old, agricultural New England aside, not to mention huge tracts of forest land, Weston is far from rural. It is affluent suburbia, home to a number of CEOs and celebrities, one of the wealthiest communities in the country. It is a place where Fairfield County real estate agents say home sale prices average more than $900,000, and many houses on the market as of this writing have prices between $1 and $2 million.

Yes, it was rural at one time. And even as corporate American converged on the region and executives bought homes in town, Weston seemed to maintain a quieter, slower pace than the rest of bustling Fairfield County. Little, if any, changes were visible.

Noticeable changes, however, came in the mid–1990s. Many of Wall Street's newly minted millionaires heard about the town's stellar school system and that the real estate here was a bargain compared to the tony towns of Greenwich, New Canaan, Darien, and Westport. There was more land for the dollar.

So Weston became discovered. Small homes, including more than century-old farmhouses, colonials, and capes, were bought, razed, and in their places hulking, four thousand–plus-square-foot homes, derisively dubbed "McMansions," sprouted. The town's centuries-old identity began to feel the strains of wealth.

Of course, several historic districts in Weston, and the restrictions they pose, precluded the loss of local history to dollars and houses to match egos. One such district is on Lyons Plains Road, which runs alongside the west branch of the Saugatuck River. Along Lyons Plains are eighteenth and nineteenth century homes close to the road and framed by stone walls. Weston's historical roots run deep in this section, not just for its farming past but for its industrial legacy, too. A number of mills once flourished here, grinding wheat into flour, making tools and wagon wheels and tanning leather. Heading several miles north on Lyons Plains, past the Emmanuel Episcopal Church where Valley Forge residents worshiped, one veers onto Valley Forge Road. Here quaintness is slowly replaced by a rise in elevation, ledge, woods and old stone foundations. Further up one finds the anachronism of more rustic homes. This is Devil's Glenn, where the Saugatuck River starts up again, side-winds into ledge crevices and forms pools directly below the dam.

Across a small cement bridge and high above the rest of Weston sits the reservoir, which extends into Redding and Easton. Owned and operated by the Kelda subsidiary Aquarion, and before that the Bridgeport Hydraulic Company, the Saugatuck Reservoir at capacity holds twelve billion gallons of potable water. When the spring thaw comes or heavy rains persist, excess water skims the top of the dam causing the river below to stir violently. If the Samuel P. Senior Dam were to ever break, Jim Daniel once cautioned at a Weston meeting, "It would cut a mile-wide swath all the way to the Long Island Sound." A frightening scenario to all who live below the dam on Lyons Plains Road, but that fear has yet to become a major concern in Weston, as the dam in post–9/11 America in closely monitored. Still, the mere thought underscores the artificiality of the reservoir, how it goes against the natural flow of the river and landscape.

But the Saugatuck Reservoir appears to be anything but artificial. It is veritable postcard of untrammeled natural purity year-round. Trimmed with white pines, hemlocks, deciduous maples, oaks, and birch, its crystal waters each winter are transformed into a seemingly endless cover of white icing, virgin except for deer tracks that crisscross it. With the spring thaw the waters become vibrant as floes abetted by icy winds move from tiny inlets and coves to course the water heading toward the dam. By early April, before the first tree buds open, black snakes and some copperheads migrate down from the hillsides to the water. Avid anglers with water company permits begin to line the shores, the only recreational activity the water company allows near the resource. Throughout the spring, summer, and early fall, these anglers cast from the rocky shores or from the overpass bridge in Redding. For them, largemouth bass, trout, or sunfish are the only mysteries of the deep.

But in early October, when the hills of Weston, Redding, and Easton peak with orange, crimson, and yellow, a new intrigue begins. The water level drops to create the effect of low tide. This happens as millions of gallons are diverted daily, pumped through a tunnel drilled through Pop's Mountain to the Aspetuck Reservoir in nearby Easton. After a summer's drought, the drop is exacerbated. A grassy patch seen straight out from the bridge in Redding becomes a mud-caked island. Ancient stone walls, the kind that line much of Connecticut, rise up like odd breakwaters. During these low periods, the Weston Boy Scouts tour the shoreline to experience nature, archaeology and local history. They often enter from the Redding side of the reservoir and make their way along ancient paths through Trout Brook Valley in Easton. The scout leaders obtain water company permission for the field trip. But one area resident now in his eighties never would seek a permit to walk the receding water's edge. Clandestinely, he would head out into the shallow waters off Valley Forge Road each autumn to scour for remains of his grandmother's house. Each year out of sentiment and contempt for those who washed away the family homestead, he would remove one stone from the foundation and struggle with it up the hill to his car to take home. Such rock solid links to his past made the annual effort worth it.

What else lies beneath? And how can we be certain these old-time, antediluvian yarns are not just folktales spun into historical fiction? Perhaps one of the most telling comments can be found in Jim Daniel's booklet, "How Things Were in Valley Forge," published February 22, 1999, by the Weston Historical Society in association with the Weston Commission for the Arts.

> Last summer, Weston's Emergency Services squad began practicing deep water rescue diving in the reservoir. The algae count was so high that they had to orient themselves by following ancient roads winding between stone walls. With flippers on their feet, masks on their faces and oxygen tanks on their backs, whenever they came to a gate they swam up the drive and had a look at the foundations of the buildings. Next summer when the water clears up, we will arrange for somebody to bring back pictures of this historic site (p. 4).

For whatever logistical reasons, the underwater photo shoot has yet to happen. But what we do have, in fact, is far more telling—late 1800s prints gleaned from photographic glass plates. Most of the black-and-white, sepia-toned pictures are the handiwork of Charles H. D. Adelbert Whitlock. Jim Daniel insists Whitlock had a professional connection to Matthew Brady, the father of photojournalism who followed Union troops onto Civil War battlefields. Though far less dramatic, Whitlock's Valley Forge photographs are among the only remaining windows into washed-away world. And they humanize it. Mr. Daniel compares the community these photos choreograph to Glocca Morra, the fictional Scottish Highland village in the Lerner and Lowe Broadway production, *Brigadoon*. "In play as well as musical, Glocca Morra was a village which came to life once a century for twenty-four hours and then vanished for another hundred years," Mr. Daniel writes in his introduction. But Weston's Glocca Morra did not have to wait a whole century to return. Unexpectedly, it resurfaced.

2

A Personal Connection

For Gary Samuelson, one of Weston's greatest historical finds, if not a personal one, began with the most mundane of tasks—cleaning out the garage. The year was 1969, and Gary, eighteen and just out of high school, moved in with his maternal grandmother, Delia Hallock. He faced a summer of chores interspersed with free time. The garage, which had been converted from a chicken coop, was part of the Hallock family homestead on Samuelson Road, off Route 57. In the late 1800s Gary's great-grandfather Charles Samuelson had operated a farm on both sides of Route 57. The Hallock family had purchased several acres and a home for $3,000 from the Samuelsons in the 1940s. Gary's parents, Jannette Hallock and Erwin Samuelson, were neighbors and childhood sweethearts.

Gary's uncle Fred Hallock, who had been living there, had just moved to Florida and left behind a garage filled with junk. Gary spent days wading and sorting through boxes of old papers, paint cans, damp wood, cracked furniture, rusted tools and assorted garbage. He tossed it all into his truck to take to the town dump, now called the transfer station.

"While I was cleaning it out I came across this little box," he recalled during an interview (February 10, 2004), shifting his hands vertically down to make a six-inch square. "I opened it, and inside were all these glass plates. I didn't know what they were."

That is, until he held them up to the light. He saw old photographic images of houses, people fishing, children, men hunting, and people standing proudly in front of their homes. Intrigued by the images, he decided to save the box, which he stored away and forgot about for ten years.

"I didn't do anything with them until the late 1970s. I had a friend who worked at Kew Photo Labs in Norwalk, and he told me there was a guy there who was really good at developing anything."

Gary took the glass plates to Kew, where developer Leonard Pravato told him these were not negatives, but photographic positives. And, yes, he could develop them. Planning to frame them, Gary had huge prints made, fourteen by twenty inches and eleven by seventeen inches.

He found the large prints, which seemed to capture a time in the late 1800s, even more fascinating than the plates. A piece of history had opened up, but he had no idea what it was. So he sought out Jim Daniel, who had been curator for Weston Town Hall for more than thirty years. At first, Jim could only surmise the obvious: they were late nineteenth-century depictions of rural Connecticut, both agricultural and lightly industrial. But where? And who were these people? Who took the photographs?

"One day afterwards I was visiting Anson Morton, who was married to my great Aunt Nellie Samuelson and lived in a house just below the reservoir," Gary recalled, going on to describe the Morton home as a ramshackle, rustic cottage that "didn't even have indoor plumbing in those days." But it seemed to suit Anson, a crusty old Yankee who had weathered the Great Depression. On Anson's wall Gary spotted a faded calendar from the 1930s, whose lone picture caught his attention. "In it there were these two guys fishing on a rock, a rock right below the dam, a rock that was right in from of Anson's house." It was the same image as one from the glass plate positives. The calendar also had something else Gary Samuelson needed to know, a photo credit. The mystery photographer was Charles H. D. Adelbert Whitlock.

Soon after, Gary returned with all the prints to Anson's home. "When I showed the rest of the pictures to Anson, he knew some of the houses. This made it all the more interesting to me, because nobody else had these photos," he said.

It became even more personal. One of the houses, a two-story colonial, belonged to Gary's maternal great-grandparents, Clinton and Eva Hull. Gary said he was only ten when Eva, "Grammy," died, but he can still hear her talking about the valley before the flood. "That's all she ever talked about," he said. In addition to his great-grandparents, Gary learned his grandmother Delia Hull Hallock had lived there, and that his mother, Janette, had been born in the house. He was relieved to discover this house was one of the few Valley Forge buildings to be moved, not burned, after the water company had condemned the land. Free to anyone willing to move it, the house was moved by trailer about a mile and a half to Godfrey Road East in 1937. Today it stands across from the transfer station, where the glass plates might have been trashed had young Gary been hasty.

Up until the late 1930s, Gary says, Clinton Hull had a thriving enterprise right in his own backyard, netting eels in the Saugatuck River that would be served up on dinner tables along the East Coast. But that business, like everything else in the valley, would come to an end. The Bridgeport Hydraulic Company paid a modest sum for the house and land, "barely a couple of thousand dollars for the house and twenty acres," said Gary. It was hardly consolation. "Clinton never really got over it. I was about seven or eight years old when he died, and he never got over it."

Clinton Hull had lost his home, the land, friends and neighbors, the factories and river he worked, and the vistas of which he was a part. His entire community, felled like the Saugatuck Valley's towering oaks, vanished with the plumes of acrid smoke from the homes set ablaze. Even the dead could not rest. Using steam shovels and a team of grave diggers, the BHC exhumed generations from the Valley Forge's Burr Cemetery and reburied them in a Redding graveyard. Clinton Hull would later be buried there, too.

3

A Valley Blooms

The area that would become known as Valley Forge wooed settlers long before our nation was born, years before the faintest murmurs of independence and revolution echoed throughout the colonies. In the early 1700s Welsh immigrants who had settled the Plymouth Plantation in Massachusetts headed South. Many were farmers, blacksmiths, and iron workers. The latter, according to Jim Daniel, had a trade that reached back to the late 1500s and the reign of Queen Elizabeth I. And they would come to operate what he calls "a bloomery."

"The word comes from a bloom, a hot wind, either of spring forcing blossoms out of plants or, by analogy, the hot wind of a furnace forcing steel out of common iron" (Daniel, p. 1). Quoting from the Oxford English Dictionary, he told me that a bloomery is "the first forge in an iron works where the metal, after being smelted down in a 'puddling furnace' into a 'ball' or 'lump' is then reheated and hammered into 'blooms' of steel".

These Welsh farmers and metal workers were obviously picky about where they would settle. Going from Plymouth to the Berkshire foothills in Southwestern Connecticut was an arduous trip. But the Saugatuck River Valley proved worth it.

"They [the iron workers] recognized that this area had three important things," Mr. Daniel told me in June 2000. "They were falling water to operate the bellows, abundant trees for charcoal, and bog iron. They put it all together, and they started the iron-and-steel industry here. At the time it was a technological marvel."

The late Weston historian Arthur "Jim" Hoe described to me (in a June 2000 interview) bog ore as an iron oxide that glacial movement deposited in low areas and swamps. He said the ore was plentiful throughout the colonies, and that much of it was concentrated along the Saugatuck River on the Weston-Redding border. Thus, the stage was set for a fledgling iron industry in a country that was yet to be.

According to Mr. Hoe, the first, full-fledged iron works factory was established in the Redding area in 1760 by Ephraim Sanford. He called the area Sanfordtown, and his son Oliver Sanford ran the iron works as well as a sawmill there. The area seemed to suit the Sanfords fine for nearly forty-five years, until 1805 when the river overflowed and the factory building was destroyed. Shortly after, Oliver moved the factory three miles south along the Saugatuck River to Weston. The location change may also have been due to new family ties. In his book *Weston: The Forging of a Connecticut Town,* Thomas J. Farnham writes that Oliver Sanford "eventually married Rachel Coley, the daughter of David Coley" who lived in Weston (p. 125). Sanford dubbed the new area Princetown, "after William Prince, a large property owner in the area. Sanford established his forge in an old grist- and sawmill and, using primitive methods, began the production of iron". (125).

Thomas Farnham differs with Jim Hoe on one point as to what made the Valley Forge an attractive location for iron works. Mr. Hoe said the presence of bog iron was a key factor, but Mr. Farnham believes that "much of the iron came from Salisbury or Kent," both in Litchfield County. Considering that Kent is more than seventy miles from Weston, Salisbury more than eighty, and that road and river were limited, Mr. Hoe's claim seems correct. More important, bog iron is still present in the area. Thus, the Saugatuck River Valley must have had the necessary components to launch one of Connecticut's foremost iron-producing communities. It was years later, when the community was fueling the War of 1812 effort, that raw iron was transported from Litchfield County iron mines. But early Valley Forge was not merely a factory town, Farnham writes:

> One should be careful to avoid imagining the existence of a modern industrial establishment in Valley Forge. Sanford was, in addition to being a manufacturer of iron, the proprietor of a farm. The operation of the forge depended, not on any regular schedule, but upon the demands of the farm and the whims of nature. During haying season, Sanford made little iron. The forge was likely to close down completely during the dry season of August and September. The ice and heavy rains of spring might also force the closing of the forge. (Farnham, p. 126)

The demand for rakes, hoes, hammers, and oxcart and wagon wheels kept the foundry busy as blooms of steel, which Jim Daniel describes in his booklet as twelve inches square, "received repeated blows from a giant water-powered hammer that would drive out and snip off impurities from a lump of common iron" (p. 2). The purified iron then became a new raw material for area blacksmiths who met the tool and transportation demands of area farmers.

"The first cast iron ploughs ever made in New England were then made here," Aaron Sanford, the then eighty-four-year-old great-great-grandson of the foundry's founder told the *Westporter-Herald* in 1938. "And people used to go around and collect old iron, bring it to the mill and have a plough made of it."

Historians note that the War of 1812 kept Valley Forge busy as munitions needs demanded more iron products. "During the War of 1812 [Sanford] sold iron to the United States government, but the expense of transporting the iron to New Haven, a six-day round trip, seriously cut into his profits" (Farnham, p. 126).

"Government contracts show that Yankee Weston supplied the iron for the War of 1812, the Civil War, and the Spanish War, and that before that, the same concern had supplied for the Revolutionary War," continued Aaron Sanford in the *Westporter-Herald* article, the exact date of which does not appear on the Weston Historical Society clipping. "The iron was sent to New Haven by the simple method of harnessing up a team of oxen. The trip was usually made in about three days, and lucky was the man chosen to go out to the big city with the heavy bars."

Two of Oliver Sanford's sons, Oliver C. and Levi, obtained ownership of the the iron works on May 31, 1813, according to Weston land records, and they attempted to revolutionize it. First, they changed the area's name to Valley Forge, and then they built another foundry across the river near the Easton border. In 1830, they put in a puddling furnace, "which changed pig iron into wrought iron and was the only puddling furnace in Connecticut at that time," Margaret Lylburn wrote for the *Westporter-Herald* in a piece titled "Did You Know That Weston Was a Big Iron-Ore Town?" (The exact date of this historical society clipping is also unknown.)

"From then on a large business in agricultural machinery and hatters' supplies was carried on," Miss Lylburn wrote. "The oxen had been replaced by horse teams; things were speeding up in old Weston! Marion Sanford tells of the government contracts which demanded that the iron ore be delivered as far away as New London."

So successful were the Sanford sons that they sold one-third of their interest in the "forge, coal house, tools and apparatus and water privilege to Morris Bradley for $500 February 21, 1832," Jim Hoe remembered. Bradley became a huge success manufacturing axes. In *Easton: Its History*, Helen Partridge writes that "Easton farmers worked (in Valley Forge) on off seasons, as well as at a pin factory below the present dam at Devil's Mouth and at Bradley's Axe factory, which turned out especially fine axes" (p. 96). The Sanfords also sold the second foundry across the river in 1840 to Bradley Hull, who quickly prospered making wagon wheel hubs. One with Hull's name on it is currently on display at Weston Town Hall.

By 1841, and eager to retire, Oliver C. Sanford turned the iron works business over to his son, William O. Sanford, who worked it until January 1, 1858, when he sold the business to a group of men and women for $3,250. The list of the fourteen investors reads like a local map, since the surnames are the appellations of many local roads today: Aaron Buckley, Nehemiah Buckley, David Siliman, George Buckley, Joseph Davis Jr., John R. Sturges, Charles Bradley, Sally Fanton, Wakeman Bradley, John Hull, Elinor Rowland, Hanford Beers, Walter Treadwell,

Fig. 6. A rustic log bridge once allowed workers of the old Bradley Axe Factory to cross Valley Forge glen. Courtesy Weston Historical Society, Inc.

and Stevan Adams. With money behind it and managed by Benjamin Franklin Buckley, the mill soon found an eager customer, the fledgling New York–New Haven Railroad. "On March 27, 1865, the N.Y., N.H. R.R. bought 1,181 pounds of crowbars—so Weston helped build up the country," Margaret Lylburn wrote.

Benjamin Franklin Buckley had a long run at the foundry managing the iron works' day-to-day operations until he died on January 13, 1900. Across the river Bradley Hull, too, worked right until his death in 1886. His nephew Henry B. Wheeler took over the factory and, appropriately, the Wheeler name appeared stamped on the region's wagon wheels. He worked until his death in 1912.

In retrospect, these deaths signaled the end of a 150-year-old way of life in the Saugatuck River Valley. In the early 1900s, most of the country's iron-and-steel industry had headed West, the same way Connecticut's strong agricultural base had pulled up stakes in the mid–1800s. After all, how could Weston compete with the likes of Pittsburgh or the iron factories in the Great Lakes region? No longer a major player in

Fig. 7. Valley Forge's Buckley homestead set on 76 acres was seized by the water company after protracted court battles and destroyed as the area was flooded.
Courtesy Weston Historical Society, Inc.

an industry forging a new nation, Valley Forge settled into the mode of a less hurried, backwoods valley, where a sense of community, nonetheless, was as tough as steel. It was this transition period that Charles Whitlock captured around 1888 on the glass plates that Gary Samuelson found.

4

A Lens on the Past

"They were either moved or burned," said Burry Fredrik, a Broadway producer and longtime Weston resident who worked with Jim Daniel when chronicling the story of Valley Forge. "That's what the hydraulic company would do. They would set them on fire because they didn't want the wood to float to the surface."

Her words were nothing if not painful when viewing the late 1800s Valley Forge photos taken by Whitlock, not to mention other unattributed class photos taken during the 1911–1912 academic year outside the Valley Forge School. They evoke melancholy because they portray daily life in a community that would later become a literal village of the dammed. They put faces on the casualties of eminent domain. In ethereal black and white, these people show a pride of home ownership, a love of place, the satisfaction of hard work and a belief that their terra firma toils would last for generations—though their descendants would have their hopes rifled.

What follows is a look at some of the Valley Forge photographs retrieved by Gary Samuelson, plus paintings, illustrations, and rare snapshots held by the Weston Historical Society. Most depict the proverbial calm before the storm, a homey, unsuspecting way of life, the days of which are numbered.

A particular Whitlock photo that touches Mr. Daniel shows a mod-

Fig. 8. A Valley Forge farm family circa late 1800, proudly displaying a white cow. The house was burned and the area is now 100 feet under water. Whitlock plate no. 10, photo courtesy of Gary Samuelson.

est, two-story farmhouse bordered by a white rail fence. A stoic yet mettlesome man sporting a frock coat stands in front. He is flanked by his wife and daughter, both wearing long, corseted dresses. More relaxed is the mustachioed man with an open collar and suspenders. He casually leans on the fence, his hat cocked to the side. Mr. Daniel figures this man is either a family friend or a farmhand. In front of the fence and nearly blocking one of the hitching posts is a boy of about ten years old. He is wearing a dark suit and a flat-brimmed hat and holding a leash to a snow-white milk cow with a full udder. A cow such as this one was a status symbol in rural America.

"Anybody would be proud of a cow like that," said Mr. Daniel. "Obviously, it was groomed for the photos."

"Bossy is so gentle," Mr. Daniel writes, "she does not even need a halter; a rope looped around her horns is enough. The lovely cow has just 'come in' (birthed her calf and resume lactating) which means there was cream on this lad's morning porridge."

Perhaps all that remains of that scene now is a leaning, algae-laden hitching post and the fieldstone foundation where fish feed.

Small-town America is also showcased in another photo already men-

Fig. 9. Dr. Frank Gorham poses outside a Valley Forge home after delivering a baby, circa 1882. The house was moved prior to the valley's flooding. Whitlock plate no. 6, photo courtesy of Gary Samuelson.

tioned, glass plate no. 6 (fig. 9), the one to which Gary Samuelson found a long family connection. It captures dignified and dapper Dr. Frank Gorham sporting a straw hat, his chest puffed out, his arm at ease as he sits perched in his horse-drawn buggy stopped in front of a Valley Forge home. The house is side-hall entry, Federal style circa 1830. In the picture there is the faint image of a woman peeking out of the second-floor window. The story behind the picture is that Gorham had just delivered a baby at the house.

> Today, Charles Whitlock has just dropped by his sister Mrs. Bradley Hull's house to see if Sarah Adelle and the baby Bradley Clinton Hull may come out into the sunlight for little Brad's first picture. We suppose this is 1882 when the baby was born. Valley Forge's Dr. Gorham has just made a house call and is about to leave. He wears a British white straw "boater," a hat style adopted for summer by American men in the latter decades of the 19th Century and worn by older men up to World War II. Because no picture of mother and child survives from

Fig. 10. Students and their school mistress, Florence Banks, gather for a class picture outside the Valley Forge School, circa 1910. The building was destroyed and the school district officially abolished when the reservoir was built. Courtesy Weston Historical Society, Inc.

this occasion, we suppose the doctor has said not this day. So he would not have lugged along his camera in vain, Charles Whitlock has prevailed upon Dr. Gorham to pose himself. In the middle upstairs bedroom window we see dimly the face of a woman looking out to see what has delayed the doctor's departure. (Daniel p. 5)

The baby, Bradley Clinton Hull, called Clinton by his parents, is Gary Samuelson's maternal great-grandfather. He married Eva Jane Williams, who had previously been married to Henry Hallock and had three children, Fred, Frank, and Eva. Fred later married Delia Stuart. The two lived in the house with their mother and stepfather, and in 1929 Delia gave birth to daughter, Janette, Gary's mother. Gary Samuelson reiterated that he feels lucky this house was one of the few saved.

According to Weston Historical Society archives, Dr. Gorham was the son of George M. Gorham, owner of a country store in Valley Forge.

Fig. 11. A close-up of the Valley Forge school children. Courtesy Weston Historical Society, Inc.

Born in 1852, Frank was raised in Valley Forge, and in 1876 was graduated from Yale Medical College. Afterward he returned to Weston to practice medicine, and he married local resident Fannie Salamon. His medical service to the community was only interrupted by a brief period when he was elected to the state legislature. An undated archive clipping from an unknown newspaper but with Margaret Lylburn's byline contains remembrances of Dr. Gorham by Frank Banks, then one of the town's oldest residents.

"Oh, yes, he was an old friend of mine," Mr. Banks told Ms. Lylburn. "I se'ed him just a few days before he died in 1926. He was only seventy-four then. The first bone he ever set was when I broke my leg over fifty years ago. Yes, that was the first bone Dr. Gorham ever set after he started practice."

In the same article, Mr. Banks described Dr. Banks as a man who enjoyed "a little chewing tobacco and a dash of whiskey for disinfecting himself." His fee for delivering a baby was twenty-five dollars.

Fig. 12. This old Colonial was one of the few Valley Forge homes that survived. It was moved to Goodhill Road. Whitlock plate no. 8, photo courtesy of Gary Samuelson.

Though the house in the Dr. Gorham photo was moved, many buildings, like the clapboard, gable-roofed, one-room schoolhouse that was the Valley Forge School, were destroyed. In perhaps the only two surviving school photos we see eight children of various ages in their Sunday best outside the school. It appears to be late spring, and the school's front door is wide open, a symbol of education opening doors. The age differences are evident from the sharp height contrasts.

The students are identified as: Fred Hallock and his younger brother Frank; Charles Rowland; Blanch Beers and her little sisters Gertrude and Elsie, each in a white dress and with a ribbon in her hair; Steve Harcor Beers; and Stanley Gould. Standing with them is teacher Florence Banks. She has deep-set eyes, short, dark hair pulled back and a stoic expression. Miss Banks is dressed in a long skirt and white blouse buttoned all the way up to her chin, and, with nary a smile, she seems the quintessential school marm.

Another Whitlock photo captures the prevailing unsuspecting mood

Fig. 13. This house built circa 1830 was burned by the water company. Whitlock plate no. 9, photo courtesy of Gary Samuelson.

Fig. 14. A rustic backwoods shot of Valley Forge men out for a day's hunt. This house, like many others, was destroyed, and the area flooded. Whitlock plate no. 7, photo courtesy of Gary Samuelson.

Fig. 15. A photographic glimpse inside the Buckley Machine Shop & Iron Works in the 1880s. The building was burned in the later 1930s, and the land is 100 feet under water. Whitlock plate no. 4, photo courtesy of Gary Samuelson.

in Valley Forge only a few decades before the water company battles. Mr. Daniel suggests it was taken in the back of the photographer's Valley Forge home sometime after lunch. The photo shows an unidentified man, one of the photographer's guests, relaxing in a Windsor chair perched on flat stones under a shade tree "to take in the view of their valley" (Daniel p. 2) The man, sporting a full handlebar mustache, has a relaxed smile, a somewhat languorous look complemented by the dozing dog, what Mr. Daniel described as "the old family collie," to his right.

Taken some time in the early spring, glass plate no. 8 (fig. 12) also captures a down home "afternoon in the valley." Outside a two-story, shingled, center-chimney colonial, we see four men wearing hats and one woman with a long, light colored apron over her floor-length dress. The woman stands in front of the open door, her left hand on the jamb. On the front step, which is a rough-cut, quarried flat stone, is a man caressing a white cat with a patch of dark fur on its head.

Fig. 16. Joseph Buckley's mill, photographed in the afternoon. In the distance a mill worker hews a log. Whitlock plate no. 2, photo courtesy of Gary Samuelson.

Below him is another man with a long, dark scraggly beard and wearing a wide-brimmed hat. He is holding a cup, and to the left of him is an earthenware jug on a small table in front of a cellar door. Mr. Daniel suggests the drink is hard cider taken from barrels stored below. Two men toward the front and behind the wood slat fence and gate pose for the camera. The one to the right wears a hat, an overcoat, and a tie. The man on the left has an open collar and is holding a light-colored pitcher. Mr. Daniel describes him as a visitor who is walking out to refill Mr. Whitlock's cider mug.

Gary Samuelson believes the visitor is the Iron Master (Franklin Buckley, owner of F. Buckley Machine Shop and Iron Works) himself. Since this man is in several pictures as a bystander, it is possible he encouraged Whitlock to take pictures to show to some banker in a nearby

Fig. 17. Two unidentified Valley Forge men in their Sunday best out for a day's hunting. This area is now beneath the Saugatuck Reservoir. Whitlock plate no. 12, photo courtesy of Gary Samuelson.

city in order to obtain a loan to repair the Buckley Iron Works after the (January 1888) blizzard (Daniel, p. 7).

According to Gary Samuelson, this is one of the few houses that survived. It was moved a couple of miles away to Goodhill Road near the intersection of Cartbridge Road. Though the house has been substantially renovated, Mr. Samuelson points out that the original architectural details are still evident.

Other homes, such as the two-story Greek Revival house that Gary found on glass plate no. 9 (fig. 13), had no such happy ending. It is assumed that the circa 1840 house with a lattice-covered well in its front yard, a house Jim Daniel describes as in "considerable disrepair," met a fate of fire and water. The same is said the of very rustic, time-worn, one-story, shingled dwelling found on glass plate no. 7 (fig.14), which shows seven men, five of them holding long-barreled rifles, out for a day's hunt in the valley with three hounds. One of the rifles, the one

Fig. 18. The photograph of men fishing that led Gary Samuelson to discover the identity of the mystery photographer. Whitlock plate no. 5, photo courtesy of Gary Samuelson.

held by the man closest to the front, appears to be a black-powder firearm. The gray-bearded man holds a double-barreled shotgun pointed skyward, the butt resting on his knee.

Mr. Daniel notes that the dwelling's rough-hewn backwoods quality "strongly suggests that before a modern iron works and foundry were built up in Valley Forge there were simple 'backyard forges'" (p. 6)

An inside look at a professional foundry, perhaps the Buckley Machine Shop and Iron Works, is seen in glass plate no. 4 (fig. 15). Pictured are the blacksmith and Mr. Buckley demonstrating, "how the hammering of iron into finished steel was done" (Daniel, p. 4). He adds that this photograph is probably the only one "anywhere today so graphically demonstrating how from the time of Elizabeth I forward an iron works modeled on the British example took 'lumps' of iron from a puddling furnace and by holding them to an anvil hammered to one side their impurities" (p. 3).

"The funny thing about all of these pictures," Gary Samuelson told me, "is that they were all taken before any of this controversy [about flooding the valley] started."Again, the people seem all so unsuspecting as they went about their daily lives. But by the early 1900s, the Bridgeport Hydraulic Company already had designs on the Saugatuck River Valley. Mr. Hoe's research shows that as early as August 25, 1914, the water company had purchased the Wheeler foundry.

Fifteen years later, as the Great Depression swept the nation and Weston was ranked one of the poorest Connecticut communities, all of the hardscrabble property owners of Valley Forge must have seemed like anxious sellers. What the water company did not expect in this forsaken valley was something that had already been forged for centuries—a love of place and an iron will.

5

A Water Company Rises

As the Saugatuck Valley slumbered, transformed from a mini-rust belt to a subsistence, backwoods hamlet during the Depression, a city fifteen miles southeast on the Long Island Sound—Bridgeport—was booming. Dubbed the Park City, or the city that the late nineteenth-century showman Phineas T. Barnum built, Bridgeport, despite the Depression, was an industrial powerhouse. It had been prospering since World War I, when its factories had been humming overtime. By the 1930s Bridgeport boasted nearly five hundred manufacturing firms that included numerous tool and die makers, cigar manufacturers, hat makers, Sikorsky Aircraft (which produced seaplanes), Bridgeport Brass, gun manufacturers, toy makers, and textile mills. These companies also fueled the city's shipping and freight rail industries.

The smoke of "progress" puffing from Bridgeport's stacks along with the rainbow glitter of industrial discharge into the Sound were 1930s symbols of success that beckoned a multiethnic swell of families anxious to find any type of work. It was during these years that Bridgeport's population rose to about 140,000, making it the most populous city in Connecticut. Surrounding towns and cities, such as Shelton, Stratford, and Fairfield, also experienced increases. And as

Fig. 19. The Joseph Buckley Forge is on the left, and the H. Wheeler Foundry is to the right of the old dam for falling water to run the bellows. The Whitlock house is on the hill. Courtesy Weston Historical Society, Inc.

the greater Bridgport area expanded, officials at the Bridgeport Hydraulic Company planned for more growth opportunities, too.

The Bridgeport Hydraulic Company had come a long way from its shaky beginnings in 1857 when it bought out the four-year-old Bridgeport Water Company, which had gone bankrupt. The Bridgeport Water Company had been formed in 1853 by Nathaniel Green in response to a major fire in 1845. The firefighters, who would pump water from the bay and harbor, were unable to maintain the flow because the tide was out, according to the historian Jim Hoe. Seeing the need, Nathaniel Green built a reservoir, "actually a large concrete tank," in the city, and he had water mains installed under Bridgeport's streets. He also built a small dam on the Pequonnock River. But Mr. Green's ambitions pushed him to take on too many projects too soon, and the result was bankruptcy.

The new investor-owned Bridgeport Hydraulic Company expanded more carefully. It grew as its customer base rose with an increased population following the Civil War. The company also seems to have bene-

Fig. 20. Another inside view of the Buckley mill. Courtesy Weston Historical Society, Inc.

fitted from its relationship with the larger-than-life business figure who in 1875 became Bridgeport's mayor—P. T. Barnum. Barnum at first suggested that the company sell itself to the city, Hoe writes, an offer the company turned down. "Because he was a good mayor or because he complained so much," Mr. Hoe believed, Barnum was named the president of BHC in 1877, a post he held for nine years, before returning to the circus. The showman to whom the phrase "There's a sucker born every minute" is attributed (though Barnum scholars say a circus competitor coined it) seemed to have infused his entrepreneurial spirit into his water company post. By 1900 Bridgeport Hydraulic had purchased and constructed fourteen reservoirs and two district tanks, an ample infrastructure for a city poised to expand its industrial and resident base with the World War I effort.

Following the war, a new president, Samuel P. Senior, took the helm at Bridgeport Hydraulic. It was a post he would keep for thirty-five years. Mr. Senior had big plans, ones that would make his water company dominate, eclipsing the Stamford Water Company and the Norwalk Water Company. He was determined to do it by building the biggest reservoir Connecticut had ever seen, and for that he set his sights on the Saugatuck River Valley.

The valley had been on the BHC's expansion agenda ever since the company first purchased the Wheeler Foundry in 1914. Other sales throughout the years were less obvious. These were watershed parcels not in the area to be flooded but needed as buffers to keep the water pure. In water company vernacular these are termed Class II Watershed Lands. Beyond them and even farther from the area to be flooded were the Class III Watershed Lands, not land necessary to make a reservoir or to keep its water pure, but land that was connected to Class II lands that the water company did not care to subdivide. By 1935, according to Mr. Hoe's research, the water company had amassed as much as 4,500 acres in and around the valley and in Easton. By 1938 the company had already secured all the rights of way to proceed with the new reservoir.

In a 1938 *Forum* article titled "Connecticut Spare My Land," author Roger Burlingame decried the way the land purchases were made. He also called to task the Connecticut Yankees, who, while revered for their legendary shrewdness and stalwart nature, sold out nonetheless.

> Propaganda is circulated over a countryside to the effect that the company has its eye on a certain valley for a reservoir. There's no specific statement of when or exactly where any development may be made. This propaganda is allowed to simmer for many years. Finally, an agent is sent to interview property owners. He makes an offer for the land. When it is refused, he says, "Well, the land will be condemned anyway; you will get a much lower price then." The absence of logic in this has not been apparent to many Connecticut Yankees, in spite of their native shrewdness. A lot of them saw the cards stacked against them and sold. Under these conditions land went for as low as $75 an acre when nearby property, just beyond the reservoir area, was sold on the open market for $1,000.

Mr. Burlingame, who claims the land purchases were virtually clandestine and to the water company's benefit, worried "that in decades to come, if the reservoir is abandoned as it is deemed no longer necessary, the land might be sold as lake frontage property for more than $2,000 per acre. Meanwhile, there are years of dark depression in all the reservoir area."

Fig. 21. Valley Forge dam and sawmill, now all under water. Courtesy Weston Historical Society, Inc.

That depression, however, was more of a fomenting anger as claims circulated that Valley Forge residents were being swindled. Among the water company's biggest critics was the Saugatuck Valley Association, headed by Weston resident John Orr Young, partner of the famed New York City advertising agency Young and Rubicam. Mr. Young's professional savvy was highly instrumental in getting his group's message spread. As a result, criticism of the water company grew at the state, as well as local, level. Even Connecticut Governor Wilbur Cross would later take the side of those aiming to stop the reservoir.

To stem the tide of criticism, BHC President Senior went on a defensive public relations campaign. Speaking before the Westport Rotary, reported the *Westporter-Herald,* Mr. Senior attempted to refute all of the attacks on the water company's ethics, particularly claims that the reservoir was not needed and that the BHC's method of land acquisition was unscrupulous.

"Let me say from the outset, with all the emphasis I can command, that this proposed reservoir in the Saugatuck is actually and positively needed now," he told the Rotary luncheon crowd at the Westport YMCA, "and this company proposed to recognize its duty to the public to build the dam at once, regardless of the obstacles which may be put in its way."

He went on to say that it was Bridgeport Hydraulic's public duty to have the dam completed and the reservoir filled by 1940, because "the next generation in Bridgeport will need all the water.... Such steps are never appreciated until shortages arrive and, perhaps, not even then by the average citizen."

Regarding what he considered the demonization of the water company, Senior tried to paint Bridgeport Hydraulic as a good citizen, good neighbor company:

> As a matter of fact, so many things have been said about us and our plans which are unfair and untrue, that we are beginning to get peeved. It is alright to turn the other cheek, but in time that, too, becomes raw. To hear folks talk about us, one would almost think we had horns. As a matter of fact, we are peaceful, law-abiding folks, just like yourselves. We take part in the charitable, social and business activities of the community. We do not cheat or misrepresent, and we believe on the whole we are doing a pretty fair job. Almost every aspirant for public office seems to want to climb into his job on our back. It is getting to be a popular outdoor amusement to throw bricks at any utility.... We have been accused of spoiling the natural beauty of the landscape. We believe the lake that will flood the Saugatuck Valley will be far more beautiful than the long rows of small cottages spotted along the river up through the length of the valley. Can anyone view a single one of the company's reservoirs without realizing that the beauty of the landscape has been enhanced? (*Westporter-Herald,* October 29, 1937)

But the fact that the company was creating waterfront properties and had already abandoned three of its reservoirs made many fear Bridgeport Hydraulic had the ulterior motive of going into the real estate business. But Senior maintained the contrary—"I would also like to say just a word in concluding about our real estate operations," he stressed. "We have never bought any land whatsoever for the purpose of speculation."

His talk must have been a hard sell since claims had circulated for years that Valley Forge residents were being cheated. True, during the

Fig. 22. Water wheel outside the Valley Forge sawmill. Courtesy Weston Historical Society, Inc.

Depression's desperate times fire sale sell-offs were commonplace. Still, residents felt they were not getting fair market prices even in hard times. A few months earlier, a Sunday *Herald Magazine* article titled "The Battle of Valley Forge" openly assailed the BHC's methods for acquiring land.

> When the Tennessee Valley Authority was forced to move some of the Tennessee hillbillies from their homes to make room for the vast body of water that was to come, they went about it with extreme care. In addition to finding better homes for the half-starved poor whites who were eeking out a precarious existence from their rocky acres, the federal authorities moved cemeteries, churches and anything else the natives thought they'd care to have. They paid, of course, a good price for

the practically valueless land. How different the attitude of the Connecticut water companies towards the Connecticut natives whose properties chance to lie in the path of their expansion activities. According to those who live in Valley Forge, Weston, and Redding, the Bridgeport Hydraulic Company intends to inundate much of the land in that region at a price which, for all practical purposes, will be set in a closed market. Although the matter of price is of vital interest to the residents, they are equally concerned with the drowning of some of Connecticut's most famed beauty spots. . . . Firm in the belief that their ancestral homes were unsurpassed for beauty and historical associations, the original natives have seen the value of their properties rise and soar as the back to the country movement has gathered momentum. Now they claim that their hopes for a fine profit are seriously menaced by the hydraulic company's threat. They point to the fact that land values have already tumbled, to the fact that the hydraulic company will be the only bidder at the proceedings because of that threat. (*Herald Magazine*, July 11, 1937)

6

Equity Versus Inequity

Their lives, like those of their ancestors, had been planted in this river valley, and from its soil a local patriotic fervor flourished. This love of place had even spread beyond Weston to Redding, Fairfield, and other towns, where people prized the natural wonders of an Eden at their doorstep. Throughout Fairfield County, even throughout the state, those who shared the valley experience—hikers, canoeists, leafpeepers, artists, writers, ornithologists, and members of the Connecticut Garden Club—all felt a sense of ownership, emotionally, at least, to what they considered a national treasure replete with trails, ancient trees, waterfalls, farms, and historic homes.

But this emotional ownership did not have the same investment of Valley Forge residents, namely, sweat, toil, and the payments of mortgages and taxes. This last requirement for literal ownership, taxes, was something Valley Forge's old Yankees accepted as a part of life. But when they learned they were paying far more in taxes than the Bridgeport Hydraulic Company, which was encircling their land, resentment against the utility grew even stronger. At an annual Weston Town Meeting reported in the *Westporter-Herald* on October 8, 1937, resident Wood Cowan, a nationally syndicated cartoonist, gave an impassioned speech.

"Is the Bridgeport Hydraulic Company, a private utility company, to be allowed to pay at a rate less than the average citizens of this town?"

VOTERS OF WESTON—BE GOOD HORSE TRADERS

The Bridgeport Hydraulic Company offers $37,500 BONUS if we will stop our fight against their dam and the destruction of Valley Forge.

This sum is not for the roads; they HAVE TO REPLACE the roads under the law. This sum is a bribe to keep this fight from going to the United States Supreme Court.

The Fight Won't Cost The Town of Weston One Cent!

The Hydraulic has already taken it to court to scare us. The Saugatuck Valley Association has guaranteed the court costs and posted a bond for it.

The Dam Won't Lower Your Taxes!

The Hydraulic owns one-third of the town of Easton. They pay one-third of the taxes — a rate kept low by law. When Easton's taxes are raised, as they will be, the Hydraulic's taxes will not be. The town takes the rap. So would we.

They Offer One Million Gallons a Day

Any competent engineer will tell you that is equal to a stream 18 inches wide, 3 inches deep. Don't be fooled by their indefinite promises.

The Hydraulic Company has no proof of their necessity for this water!

We have a strong case. They know it. That's why they are trying to buy us off on the eve of the first court hearing on this necessity.

Governor Cross Has Promised Us His Aid!

Wait and see what he can do. If you sell out, you've sold your rights forever — for $37,500.

Let's Remember We Are Americans!

Don't let a privately owned utility company be dictator in our town! God gave us the water. Why should we let it be taken to Bridgeport to enrich the Hydraulic stockholders?

Come to Town Meeting Wednesday Night! Vote Against This Bribe!

VOTE NO!

Fig. 23. A copy of a Saugatuck Valley Association poster circa 1937, reading "Voters of Weston Be Good Horse Traders." Courtesy Weston Historical Society, Inc.

Mr. Cowan declared, "Mrs. Lydia Sokoloff is assessed at a rate of $153 an acre. She is entirely surrounded by the water company property, which is assessed at $56.60 per acre. Are we as citizens of this town going to stand for this?"

Cowan charged that while the Bridgeport Hydraulic Company owned one-sixth of Weston's land mass as of 1937, "it pays one-nineteenth of the taxes." Mr. Cowan gave myriad examples.

"Jimmie Griffin, whose land the water company is now trying to condemn, has his land assessed at a rate of $81 an acre," Mr. Cowan stated. "Jimmie is entirely surrounded by the water company which pays a rate of $56 plus. How does a thing like this happen?"

Mr. Cowan also accused the water company of using two sets of valuation for its Weston properties. One, for the purpose of the town's tax list, had its properties valued at $121,000, he said, whereas another for the purpose of its water rates put a value of $590,000 on its real estate holdings. "The president of the hydraulic company on the witness stand the other day testified that its holdings amounted to a half-a-million dollars."

Based upon Mr. Cowan's claims, the article went on to what a number of Weston residents were paying in property taxes per acre compared to the BHC's $56.60 per acre taxes. "Franklin P. Adams pays at a rate of $136 plus; Mrs. John Hull Brewster at $100; Mrs. Frank Cobb at $165; James Daugherty at $168 plus; Phillip Dunning at $195 plus; Cecil Holm at $100; Lee Keedick at $150 plus; William Meade Prince at $155; Edgar Perry at $100 plus; Emma Bradley at $122 plus; David S. Coley at $140; Frank Fitch at $150; Douglas Haddan, at $217 plus." The exhaustive list proves Mr. Cowan's argument. Citing a Connecticut law stating that only land seized by a municipality for its own use would be tax exempt, he pointed out that other lands, such as those seized by Bridgeport Hydraulic, were to be taxed at the same rate as farmland within the town. "The intelligence of this town is high above average," he said. "Are we going to allow a private corporation such privileges over our private citizens?"

Mr. Cowan's arguments were highly persuasive, carrying the necessary pathos and reason to sway the audience. Not that he had to try hard. He was, after all, preaching to the proverbial choir, a Weston audience. Town officials voted unanimously to increase the water company's tax bill.

But corrective action against tax inequity was a minor skirmish compared to mounting concern that Valley Forge residents had been cheated out of a fair price when being forced to sell to the water company.

> In addition to the natural beauty that the hydraulic company proposes to hide under the placid covering of a sheet of water are the fine old Colonial homes that stand sturdily today in that section. Built with bitter hardship and loving care, those old homes express the very spirit of the hardy settlers and are the state's best possible links with its historic past. They are treasured alike both by descendants of the original builders and by the ever increasing army of Americans who have joined the "back to nature" army. This last fact, present land owners vehemently claim, is what is back of the hydraulic company's condemnation proceedings. They see the move as outright speculation in land and point to the fact that water company officials have admitted that they may not flood the section as proof of suspicion the whole thing is based on expectation of realty profits. There is no doubt that the condemnation of this section would leave many natives holding a bag they have tightly clutched down through the years. . . . Students of Connecticut's so-called public utilities will recognize in this fight the old, old question of just what a utility's duty is to the citizens in the state which supports it. (*Herald Magazine*, July 11, 1937)

That question eventually found its way to Bridgeport Superior Court where a board of appraisers appointed by Judge Kenneth Wynne publicly heard arguments about Valley Forge land values. A *Westporter-Herald* front-page article, "Prices Paid for Land by the Hydraulic Co. Are Aired in Court," published November 2, 1937, described debate over two Weston parcels. The first properties under appraisal were ninety-one acres belonging to James Griffin and seventy-six acres owned by Edgar Perry. During the hearings, BHC President Senior admitted he placed the same valuation on land set back deep into the woods as those fronting improved roads.

"He admitted that the company had paid $16,000 to one property owner for three acres, but that he considered Mr. Griffin's ninety-one acres worth only $17,450 and Mr. Perry's 70 worth but $14,040," the article states.

Speaking on behalf of the landowners was Daniel R. Harvey, Weston's third selectman, who said the properties should be worth much more since they fronted a new Valley Forge Road constructed in 1936. Further, the road had been built "despite a letter from Mr. Senior advising against its construction."

"Property values were materially increased," Mr. Harvey stated, adding that prior to the new road Valley Forge was virtually impassable to cars and school buses for two to four weeks during the spring thaw mud season.

After Mr. Harvey spoke, Mr. Senior took the stand, where he adamantly responded, "I think it is a lot of bunk that road increases property valuations in the valley." According to the article, Mr. Senior went onto to testify that over the past five years Bridgeport Hydraulic had paid the following fair prices for Valley Forge properties: "Oct. 30, 1937, H. Stanley Todd, Redding, $6,500 for 96 acres; Oct. 1937, Theresa McMahon on Aspetuck River, $2,000 for 15 acres; Fred Driggs, Redding, $3,000 for 30 acres; E. H. Delafield, Weston, $4,000 for 30 acres." According to Mr. Senior's reports, prices paid a year earlier were even lower: "Frank Morehouse, 11 acres in Weston, $500; May 1936, Carrie and Charles Morehouse, 33 acres in Weston, $1,500; August 1934, Lloyd and Blackman, 72.5 acres in Redding, $7,250; February 1934, J. E. Godfrey, 19 acres in Weston, $1,400; Wesley Norton in Easton, 10 acres, $450."

The Griffin-Perry property hearings, which had been going on for several days, even included testimony from a prominent Fairfield County real estate agent, Walter Peck, who took "the stand in favor of the two 'condemned' men," (*Westporter-Herald*, October 31, 1937).

"Walter Peck, who said he had twenty-five years of experience in the real estate field, was cross-examined at length by the attorney for the hydraulic company who endeavored to shake Mr. Peck's testimony," the article states. Mr. Peck had appraised the land in question at $300 per acre.

"Even these three or four acres of swamp?" the unnamed water company examiner reportedly countered.

"Yes," Mr. Peck responded without hesitation. "When I buy a porterhouse steak—not that I can afford one often—I know that no matter how good the rest of the steak, there's always a little tail. That's the way I've found real estate—you can't help getting some tail."

Despite these efforts, the board of appraisers sided with the water company, which should not have come as a surprise. The board's members—Frank L. Wilder of Bridgeport, George Van Riper of Westport, and Samuel Keeler of Wilton—had been appointed by Judge Wynne, who earlier that month had ruled in favor of the hydraulic company's petition to condemn the Griffin and Perry properties in Valley Forge, as well as the property of Duncan Gay in Redding Glen.

"The only issue in these cases is whether the undoubted power of the applicant (BHC) has been exercised properly or abused," Judge Wynne was quoted saying to the *Westporter-Herald* on October 8, 1937. "The policy of the state in reference to the taking of property in the public interest is definitely fixed and has been approved judicially so many times as to need no comment here."

To Weston landowners who felt they had been bulldozed by the water company, Judge Wynne's comments and the resulting rulings fueled their cynicism. Some filed suits arguing their lands were worth much more than the water company was paying. But their suits against a Bridgeport utility were being decided in a Bridgeport court by Bridgeport judges. What more could they expect?

Exacerbating their feelings of being cheated was the fact that another eminent domain buyout was simultaneously underway by the State of Connecticut then poised to construct the Merritt Parkway, the state's first major highway. Payments from the state were exceedingly, if not ridiculously, generous. And the rub for Weston residents was that many of these parkway lands were less than a mile from the Weston border.

"These sales have been under consummation for a long time, and undoubtedly the State of Connecticut must have been making these purchases with its eyes wide open," states a front-page *Westporter-Herald* article, "High Prices Paid for Parkway Land Cause Stir," which ran on Christmas Eve, 1937.

The article could not have been more timely. The state came off looking like Santa Claus. The most extreme example is that of Mrs. John

Cavanaugh of South Norwalk whose 14.8 acres had been assessed at $3,900. The state purchased the parcel for $57,000. Lucie Woolson's 11.4 acres valued at $1,149 was bought for $14,558. Mary M. Inman sold her 24.1 acres assessed at $6,900 to the state for $33,808, and Rachel M. Taylor's 14.4 acres assessed at $7,350 netted $22,500. So for the struggling Depression-era farmers and landowners along the planned Merritt Parkway, eminent domain proved a windfall.

But Valley Forge residents would not even come close to their lands' assessed values, let alone a windfall. They would be forced to take what the water company offered, and since BHC focused on the bottom line, the offers were low.

7

The Fighters

Lillian Wald was a fighter eager to battle poverty, hunger, disease and other ills that had a stranglehold on New York City's slums at the turn of the last century. A literal foreshadowing of beatified Mother Teresa of Calcutta, who would draw worldwide attention nearly a century later, Lillian Wald ministered the downtrodden, the destitute, all those deemed human refuse by an industrial society hardened to any kind of compassion. Trained as a nurse, Mrs. Wald in 1893 opened the Henry Street Settlement in lower Manhattan. There she taught impoverished immigrant women nursing skills and pushed for programs offering better hygiene to the poor. She also expanded her settlement house, which later became known as the Visiting Nurse Society, to include social services. For the disenfranchised she established a library and a savings bank and instituted vocational training programs for children. An advocate for child labor laws and women's rights, she founded the Women's Trade Union League and the National Child Labor Committee. She was also the founder of Columbia University's School of Nursing. A tireless humanitarian, Mrs. Wald was an outspoken pacifist protesting the United States' involvement in World War I. She became a member of the International League for Peace and Freedom and the Women's Peace Party. She is revered as one of the great social reformers of the twentieth-century.

SPECIAL TO WESTON

You people of Weston want facts! — facts on which to base your answer to one of the most serious questions you as citizens of this Town will ever be called upon to decide — WILL YOU ACCEPT THE DAM WITHOUT A STRUGGLE OR WILL YOU FIGHT FOR YOUR RIGHTS?

At a special Town meeting, Thursday evening, March 10, the Town voted to rescind the previous resolution of firm opposition to the proposed reservoir. A special committee was appointed to investigate the activities of the Bridgeport Hydraulic Company in this matter.

BUT WHAT HAPPENED? Bright and early Friday morning the water company unexpectedly brought suit against the town to take over the roads. The corporation disregarded completely the Town's friendly gesture of the evening before. The corporation also chose to overlook the Town's appointment of a special "investigation committee." They greeted your truce with a volley of shots!

AND HERE IS WHERE YOU STAND TODAY! The Town is faced with a lawsuit — a suit in which the Bridgeport Hydraulic Company will do all in its power to squeeze the last dollar it can from the Town. AND WHAT ARE YOU DOING? You are sending your attorneys into this battle minus one of the strongest weapons you can give them— the Town's concerted, uncompromising opposition to construction of the dam!

EVERYONE IN WESTON IS ASKING QUESTIONS! HERE ARE SOME OF THEM:

"WILL OUR TAXES BE RAISED TO PAY THE LEGAL EXPENSE OF FIGHTING THE SUIT?"

Absolutely not!

"WHAT CAUSED THE RECENT INCREASE IN OUR TAX ASSESSMENTS?"

A great deal of confusion has resulted from the recent blanket increase in every one's tax assessment in Weston. An unfounded rumor has been going the rounds that the Saugatuck Valley Association is responsible for this increase and that the purpose of the increase is to provide money to pay for the expense of the lawsuit. There's no truth in this.

At a Town meeting last Fall ONE of the Town's voters — NOT THE SAUGATUCK VALLEY ASSOCIATION — eloquently championed the idea of increasing the assessment on land held by BHC. A resolution was prepared and the Town voted affirmatively on it. BHC, however, would have none of it and by law forced the Board of Assessors to raise every one's assessment proportionately. A reduction of the mill rate straightened things out and every one ended up by paying the same taxes he would have paid if no jockeying had occurred.

"HOW MUCH WILL THE SUIT COST?"

Not more than $2500 to carry it through the Superior Court. In the event of an appeal by either party, an additional $2500 will be more than ample to see the Town through the Supreme Court of Errors.

"HOW WILL WESTON PEOPLE SUFFER IF THE DAM IS BUILT?"

If the proposed reservoir project is not defeated, more than 18% of the land of our town will be removed forever from normal development. Weston will become just another of those dead communities labelled "reservoir town." DEAD because the heart would be cut out of it. IS THIS WHAT YOU WANT?

"HOW ABOUT THE PROPERTY OWNERS ABOVE THE DAM?"

You would cease to be a free citizen! BHC would tell you what you can and cannot do on your own property. People in Easton attempting to live under these conditions describe the situation as "unbearable."

"HOW ABOUT THE PROPERTY OWNERS BELOW THE DAM?"

People below the dam are at the mercy of BHC—whether or not you lose your river will be decided by the whim of the board of directors of this public utility.

Oral "promises" have been made that the minimum flow will be maintained at all times. But what is meant by "minimum flow"? And why won't BHC put this promise in writing?

Mr. Senior tells us that we may expect a certain amount of water from seepage. "Seepage" from a concrete masonry dam bedded deep into virgin rock? Seepage sufficient to influence the flow of our river may mean only one thing— grab your family and run before the dam comes down around your ears!

Yes, you either stand to lose your river—or get enough of it to wipe you right off the map!

"WHAT'S THIS TALK ABOUT A DAM ON THE WEST BRANCH?"

One of BHC's top executives stated, "After we take the Saugatuck, then we'll dam up the West Branch."

Just like that this "private" utility wantonly condemns another valuable piece of Weston to death!

"WHY DO ANYTHING ABOUT IT NOW?"

Because NOW is the best time to stop it! Why wait for 15 or 20 years just to hear the old defeatist slogan, "You're 20 years too late!"?

"IF THE DAM GOES IN, WHAT TAXES MIGHT WE EXPECT FROM BHC?"

AT BEST THE BRIDGEPORT HYDRAULIC COMPANY WOULD NEVER EVEN PAY ITS OWN SHARE OF WESTON'S TAXES! The water company already owns, controls, and restricts the development of more than 18% of all the assessable land in Weston. And yet, if the dam were built, the corporation would never contribute more than 15% of the Town's total income from taxes. And through the years, as the land in Weston becomes more and more valuable and the Grand List of Weston mounts, the percentage contributed by the water company would become smaller and smaller. And why is this? Because, by law, land held by BHC could not be assessed for more than farmland — and the assessment on the dam itself would remain static, perhaps depreciate!

"IF WE BLOCK THE DAM, WHAT WILL HAPPEN TO THE TAXES?"

More than 2,000 acres of some of the choicest land in Weston, which the Bridgeport Hydraulic Company has held and depressed for years, will be turned back into circulation!

Today the average acre in Weston owned by private citizens is assessed at $285. The average acre held by the BHC is assessed at $109. In blocking off this land, the BHC has thus depressed the valuation of overy acres of it by approximately $176. For the total acreage held by the corporation this amounts to a potential loss in valuation today of $400,000! Think what this land will be assessed at in the future!

WHAT IS YOUR PART IN THIS?

In this democracy of ours no intrenched wrong can withstand the impact of aroused public opinion. This Association of yours is voluntarily directed by public spirited men and women who have taken up the fight in your behalf. Surely, none of us is indifferent to the prehensile activities of the Bridgeport Hydraulic Company. You either applaud them or you abominate them. Take a firm position one way or the other. Come out in the open.

THIS IS A CHALLENGE TO YOUR FIGHTING SPIRIT!

If you are with us, say so on a postal card, by wire if you are out of town, and by coming to meetings of the Association when they are called.

Show this to your friends and neighbors. Talk about the whole subject at every opportunity. Get excited about it. And don't hesitate to ask questions when they come.

Be on the lookout for your next bulletin. Help us to help you **BLOCK THE DAM! SAVE THE SAUGATUCK!!**

Fig. 24. Original "Save the Saugatuck" poster circa 1937. Courtesy Weston Historical Society, Inc.

In 1933, at the age of sixty-six, her activism slowed when she suffered a stroke and retired to what has been called her "house on the pond in Westport." There she was visited by luminaries like Eleanor Roosevelt and Albert Einstein. Even though debilitated, Mrs. Wald kept her hand on the pulse of social issues, mustering the strength to try to right social wrongs. In November 1937 she welcomed a *Westporter-Herald* reporter to her home, where she expressed outrage about a pressing social issue in her new backyard—the possible loss of the Saugatuck River Valley.

"It is tragic to think a private utility, such as the Bridgeport Hydraulic Company, can destroy for its own profit and without showing any necessity for the public good, such a tract of natural beauty as the Saugatuck River Valley," declared Mrs. Wald to the *Herald* (November 19, 1937).

The *Herald* article describes Mrs. Wald's frail health as almost a foil to her sharp mind as she "now sits in her beautiful, comfortable, old house-on-the-pond in Westport with an illness standing between her and the world she has nursed so well. Now she takes her tea quietly before her own fireplace, and the world comes to her when it can get past the doctor. . . . as I talked to her about the Saugatuck Valley the other day I had the strange feeling that the physical self of Lillian Wald is transparent, and it is her soul who now has substance."

Lillian Wald was never a woman to hide feelings. She preferred passion to passivity, being outspoken to remaining silent.

"I know that its charter gives the hydraulic company the right to take any stream in Fairfield County, but there are so many arguments for rescinding a charter granted under other circumstances and other deeds," she told the Herald. "It is no secret that the legislature of Connecticut was once, and for a long time, controlled by boss rule for private gain of individuals and without concern for the greatest number, but those times are outdated, and a new conscience of right of the people has spread over the land."

It was not unusual to see a fire for social issues still burn in one of the century's greatest movers and shakers. But beyond the rhetoric was an acute mind in control of all the facts behind the case, an impressive grasp considering the debilitating stroke that would claim her life in less than seven years.

"During these past years of privilege the Bridgeport Hydraulic Company has built twelve reservoirs," she said. "It now uses three of them to supply water to its clients. Three of them lie idle. Six of them have been abandoned in the past ten years. Three of the six are now on the real estate market. Of the three that are in use, little more than half their daily output is used, and in the driest years they were never drawn down to one-third their safety draw down. Some of the experts who have been working on this problem say the proposed reservoir will not be needed for fifty or sixty years. Many of them say that it will never be needed."

Her statements to the *Herald* agreed with those made a month earlier by James F. Sanborn, a Westport civil engineer who was the chief engineer supporting the Saugatuck Valley Association's case. Mr. Sanborn had spent weeks poring over facts and figures to prove the reservoir was not needed. His report published in the *Westporter-Herald* (October 15, 1937) showed that the Bridgeport Hydraulic Company had built a dozen reservoirs, and only three were actually being used: Trap Falls, Easton Lake, and Hemlocks Reservoir.

"Three Shelton reservoirs were lying idle," Sanborn was quoted saying in the article "Engineer Reveals Statistics on Needs of Hydraulic Co." that "Six have been abandoned in the past ten years." He went on to state that the storage capacity of the three reservoirs in use was 12.5 billion gallons, and that the proposed reservoir would double that. Further, he said, the three reservoirs yielded about fifty million gallons a day, but that only twenty-seven million gallons per day were actually being used, half for residential and half for commercial consumption.

This was the very case state Senator J. Kenneth Bradley, a pro bono lawyer for the Saugatuck Valley Association, argued before Judge Kenneth Wynne in Bridgeport Superior Court to stop the condemnation of Duncan Gay's property in Redding Glenn and James Griffin's and Edgar Perry's land in Valley Forge. That argument, however, did not convince Judge Wynne, who ruled in favor of Bridgeport Hydraulic. Still, Sen. Bradley's rhetoric that the public welfare neither required nor justified the water company's exercise of eminent domain resonated throughout the state. And it was this sense of injustice that piqued Lillian Wald's interest to champion the Saugatuck Valley Association. Throughout Westport, Weston, Wilton, and Redding, the SVA's membership roster grew to include many ex-urbanites, specifically artists, writers, and painters, who felt an affinity for unspoiled, natural places.

"I am gratified to see that there has arisen so great a number of people who are not afraid to be articulate and to denounce the sacrifice of the present Connecticut to an enactment that could not pass today and to which the possessors of this privilege have not shown and cannot show an urgent necessity," Lillian Wald told the *Westporter-Herald*.

Then with the inimitable drive that she drew upon to champion world peace, child labor reform, and health and employment opportunities for the poor, Lillian Wald trumpeted her call for saving the valley.

"The pledge card of the Saugatuck Valley Association says that there are no obligations and no dues," she said. "But it is the obligation of everyone who knows and loves this beautiful spot not to sit quiet and see it despoiled, but to rise up and do what we can. A combined strong protest of the people will be heard not only in Connecticut but in Washington and throughout the country which is dedicated to the welfare of the many, not to the profit pursuits of a private utility."

Speaking before the Westport Chamber of Commerce, Weston writer Barton Davis gave a progress report of the SVA saying that more than one thousand people had signed pledge cards and that more were signing up daily.

"The Saugatuck Valley Association had passed the first stage of hot indignation in which everyone wanted to do something to save the river and the scenic beauty of the threatened area but didn't know what to do," Mr. Davis told the *Westporter-Herald,* October 8, 1937. He went on to say that the association actually had two missions, "one, to carry on a legal fight against the building of the dam, and the other to preserve and develop the beauty and usefulness of the river after it had been saved."

Mr. Davis also appealed to Westport residents, who, while it was not their land being condemned and flooded, would nonetheless be affected as the rushing Saugatuck River below the proposed dam ebbed. The issue, he said, was "tied up with the riparian rights of every land owner on the lower Saugatuck and with the enjoyment of the river by every Westport resident, year-round or summer, and every taxpayer."

In addition, he said that areas such as Redding Glenn and Devil's Den (later renamed Devil's Glenn after the reservoir was built) had "been compared by travelled persons and artists to the best of Colorado and the Black Forest."

In that same issue of the *Westporter-Herald* another article reported how the Federated Garden Clubs of Connecticut, comprising seventy-two local clubs and with a total membership of 4,149, threw their support behind the Saugatuck Valley Association. The state groups had been rallied by Mrs. John A. Baker, president of the Westport Garden Club. The Connecticut group's official statement read, "The Federated Garden Clubs of Connecticut, because of their interest in preserving natural beauty of rare value to the state, opposes the flooding of Redding Glen and Valley Forge until some time can be proved that it is in the public interest."

"As president of the Westport Garden Club and with the full concurrence of the executive board, I am presenting your large group of garden club members a problem in conservation which we are facing in Fairfield County at this moment," Mrs. Baker stated.

"We have in the Saugatuck River Valley some of the rarest and most arresting beauty in our country," she continued about the ancient oaks, maples, hemlocks, and mountain laurel perched on hillsides and rock outcroppings above expansive fields cut by the river. "It was early last summer when one of our garden clubs called attention of its neighbors to the need to act at once in defense of this treasure of ours. Instead of contemplating a possible distant use of this river for its potential water supply, we were actually faced with the prospect of seeing it wiped out by a reservoir."

Artists who drew inspiration from the Saugatuck River Valley brandished brushes, charcoals, watercolors, oils, easels, and cameras to raise funds for the SVA's publicity campaign and legal battles. An art show, "A Pictorial Exhibition of the Saugatuck Valley," was held at Westport's General Putnam Inn on November 12, 1939. The idea for the event originated with local artist Howard Heath, inspired after he spied sketches for the Merritt Parkway.

The exhibition opened with a silver tea, the *Westporter-Herald* reported, and included the efforts of dozens of artists. Some of the other artists were Karl Anderson, Howard Heath, George Wright, James Daugherty, Edward Boyd, George Efron, Lowell L. Balcolm and Neil Dorr.

The show was such a success that it opened again in December at the Ferragil Galleries at 63 East Fifty-Seventh Street in New York City.

SAUGATUCK VALLEY DEFENDER

Speaking for the Saugatuck Valley Association, the permanent, non-partisan organization pledged to the preservation and improvement of the Saugatuck Valley.

VOL. 1, NO. 1   "SAVE THE SAUGATUCK"   April 1938

BRIDGEPORT HYDRAULIC BEATEN BY ASSOCIATION IN COURT TILT

PLAIN FACTS WORTH KNOWING ABOUT SAUGATUCK VALLEY

WHAT IS THE SAUGATUCK VALLEY ASSOCIATION?

The Association is a voluntary organization of public-spirited men and women pledged to the protection of Fairfield County's rivers and streams from exploitation.

WHO ARE SOME OF THESE PUBLIC-SPIRITED MEN AND WOMEN?

Among those helping to manage the Association and supporting it in its work are John Orr Young, Roger Burlingame, Stuart Chase, Jascha Heifetz, Miss Rachel Crothers, Capt. Frank Hawks, Col. Edward J. Steichen, James Melton, Miss Lillian Wald, John V. N. Dorr, Karl Anderson, Prof. Raymond Rodgers, and a host of other prominent persons whose integrity and sincerity of purpose are beyond question.

WHAT IS THE ASSOCIATION DOING TODAY?

It is engaged in a vigorous, truth-seeking crusade — a people's protest against the un-American type of power wielded by one particular utility company in Fairfield County — the Bridgeport Hydraulic Company.

WHAT GIVES THE BRIDGEPORT HYDRAULIC COMPANY SUCH AMAZING POWERS?

The corporation's power rests in its loose-jointed charter which was granted to it by the State Legislature in 1857 and liberally amended in 1927.

HOW BROAD ARE THE POWERS OF THIS CHARTER?

Broad as the heavens! The Bridgeport Hydraulic Company has the power to take any land whatsoever in the whole of Fairfield County, whenever it deems it "necessary OR expedient."

WHAT DOES THIS MEAN TO YOU?

It means that this corporation—at a nod from its board of directors—can among other things seize your home and your land, build a dam in your back yard, flood thousands of acres of the best land in your town!

WHERE IS BHC'S ATTENTION FOCUSED TODAY?

On the Saugatuck Valley. BHC proposes to dam the Saugatuck River, flood the heart of Weston and Redding, conduct the river through a tunnel to Easton.

WHAT IS THE ASSOCIATION DOING ABOUT IT?

We have retained legal counsel to carry our fight into the courts. We have a corps of outstanding engineers studying the problem. We are holding meetings, committees are at work, in short, we are carrying the fight forward for you and your neighbors.

WHAT CAN YOU DO ABOUT IT?

Read the SAUGATUCK VALLEY DEFENDER — every word of it. Show it to your friends and neighbors. Talk about this vital subject. Get excited about it. And if you are having any trouble with the BHC, get in touch with us immediately — WE WANT TO HELP YOU!

Rempsen Wins Right to Appeal to Supreme Court

The Saugatuck Valley Association scored a clean cut legal victory in its fight to block the Bridgeport Hydraulic Company's plans to build a dam in the Saugatuck Valley. Judge Edward J. Quinlan entered a judgment in Superior Court in the corporation's condemnation suit against J. J. Rempsen which allows the defendant to appeal to the Supreme Court of Errors from the decision on the demurrer handed down by Superior Court recently.

The Corporation was balked in its efforts to prevent this appeal to a higher court; also in its efforts to group four other condemnation suits against property owners under the Court's previous ruling in the Rempsen case.

Despite vigorous opposition from the utility's counsel, these four cases were continued by order of the Court until the Supreme Court has rendered a decision in the Rempsen case. The defendants in these suits include Walter and Cornelia D. Hochstrasser, Marion T. Pearson, Dorothy Deyo Munro, Malcolm Curry.

Turning Point in Challenge

"This is the opportunity we have been waiting for," said John Orr Young, SVA president, "the chance to have the questions of law decided by the Supreme Court before the questions of fact are brought before the Superior Court in the other condemnation cases.

"All our previous setbacks in the courts were expected. We have never looked on them as defeats, but rather as stepping stones to an appeal to a higher court where the question of necessity will be tried. This is the turning point in the success of our challenge to a public utility which has always nurtured the belief that it is invincible. We don't believe it!" DO YOU?

WESTON MEETING

A special meeting of the Saugatuck Valley Association for the people of Weston will be held at 8:30 Saturday evening, April 9, at the Horace C. Hurlbutt, Jr. School.

Fig. 25. Original copy of the April 1938 premier issue of *The Saugatuck Valley Defender,* the newsletter of the Saugatuck Valley Association. Courtesy Weston Historical Society, Inc.

According to the *Westporter-Herald* (December 17, 1937), more than five hundred people attended the Manhattan opening, which featured even more artists. The most celebrated was John Stuart Curry whose valley-inspired painting, "Laurels," graced the program's cover. Another painting, a unnamed oil by Walter Tittle, the article states, "portrays James Griffin and Tom Bennett, two old time Connecticut residents of the Saugatuck Valley who play for square dances thereabouts. The eight-acre property of the one in the foreground (Mr. Griffin) is now in the condemnation court."

The fact that artists had entered the fight, and that the struggle was perceived as an abuse of capitalism, grabbed the attention of Connecticut's Socialist Party. Following a meeting of its executive committee, the Socialists announced their support for the Saugatuck Valley Association, stating that all should "recognize that the issue is one of public ownership and use of our natural resources, not merely a sentimental crusade of country dwellers against an acquisitive corporation."

By now the Saugatuck Valley Association had grown to more than 1,500 members. Along with teas and art shows, the group even had their own women's softball team, the Saugatuck Saviours. The Saviours held fund-raising games throughout Fairfield County. To grab press attention they even featured an occasional celebrity player, such as part-time Weston resident Gypsy Rose Lee, who pitched, according to Weston resident Patricia Heifetz.

Beyond the cause célèbre, more people miles from the valley wondered how damming the river would directly affect them. Some feared that if the water company kept growing there would be fewer houses built, which would force existing homeowners to bear a higher tax burden.

"There are three watersheds left to seize and turn into artificial lakes all neatly planted with evergreens in rows like tombstones in a military cemetery," said Barton Davis of Weston. "In this every taxpayer and resident, whether he owns land or does business in these towns or not, has a personal interest touching his or her pocketbook."

The watersheds he was referring to included the Norwalk River, the Saugatuck Valley, and another valley to the east of it just beyond Pop's Mountain, a scenic watershed that would later be known as Trout Brook Valley. Bridgeport Hydraulic had already been buying that land, too.

8

Road Wars and Water Rights

By the beginning of 1938 Samuel P. Senior was eager to get his project moving, and he exuded the confidence of someone who had the courts on his side. The Saugatuck Valley Association, the back-to-nature do-gooders and all the grandstanding artists and politicians, as he saw them, were just like the black flies that filled the valley each spring, annoying, but they would be gone soon enough. To him, the real power rested in Bridgeport, and his case was as ironclad as the company's charter steeled with eminent domain from the state.

Naturally, he knew there would be some rearranging of the infrastructure around Valley Forge. When the valley would be flooded, the roads, like the houses, trees, and everything else, would be gone. Mr. Senior wanted to make sure Weston did not give him a hard time about his company's plans for rerouting the soon-to-be-flooded roads. In early January he sent Weston's first selectman Chester Coley a letter that echoed the same bullying tone landowners had heard years earlier. In a letter published January 4, 1938, in the *Westporter-Herald*, Senior wrote,

> As you know, this company will find it necessary, in order to construct its Saugatuck Reservoir, to secure the discontinuance of existing highways for this purpose. However, as you know, if court action should be

necessary, the new layouts will be determined by the court in its best judgment and would be binding upon all concerned.

We suggest that it may seem to you desirable and advantageous to the town to discuss in advance of such court action the possibility of an agreement as a discontinuence of highways in this section as to such damages will be sustained by the town by reason of these changes. We should much prefer to make every reasonable effort to meet the wishes of the town in this matter rather than apply to the Superior Court.

Chester Coley was blunt in his response.

"We think that you should submit to use, so that we can submit to the Town Meeting, a proposal and a plan for the relocation which covers all of the points involved," he responded in a letter printed in the same issue. "As you know, there are a large number of people in Weston who are opposed, and they will probably vote that the selectmen contest without compromise any action of the hydraulic company."

Mr. Coley insisted that there were a number of points the water company had to address, including the relocation of these roads to be buried under water: Westport-Danbury Road, Godfrey Street, Davis Hill Road, Norwalk-Newtown Turnpike, Old Dimon Road, Den Road and Old Forge Road. Other matters included the construction of a new bridge to replace the one that would be lost, the appraisal of damages to property owners whose homes would be on discontinued roads, and the acquisition of the necessary rights of way. The water company, Mr. Coley continued, would also have to reimburse the town about $30,000 in state aid that had been anticipated to construct the Valley Forge Road. There were also the matters of a town inconvenience and a grand-scale interruption of traffic caused by the road relocation. "There are many other points which we have not attempted to cover in this matter," he wrote.

A Weston town meeting in March established a committee to look into water company responsibilities to Weston. The committee included First Selectman Chester Coley, Selectman Daniel Harvey, residents John Anderson, George Waterbury, and the Saugatuck Valley Association's John Orr Young and John Dorr. For months they met with water company officials to reach a compromise. Meanwhile, Bridgeport Hydraulic had already instituted legal action in Bridgeport Superior Court demanding Weston close certain roads and relocate others. By May the committee had fashioned a settlement that would ultimately face a Town Meeting vote.

Under the agreement the water company would build a paved road "of similar character to the present Valley Road" (*Westporter-Herald*, May 20, 1938). The water company also agreed to spend $5,000 on improvements to David Hill Road, and would pay the town $37,500 on or before October 1, 1938. The water company also promised to release at least one million gallons of water a day for ten years "from the date the company starts filling the reservoir and will maintain a reasonable flow of water below the dam of not less than the annual minimum flow averaged over a reasonable period of years," states the article, "Weston-Hydraulic Co. Proposed Plan of Settlement Outlined." Bridgeport Hydraulic also agreed to let town residents hunt and fish on their lands "under reasonable regulations."

Not everyone in town was thrilled by the planned agreement, nor was everyone on the committee that fashioned it. Before the Town Meeting had a chance to vote on it, committee member John Young wrote a front-page *Westporter-Herald* missive assailing the plan. In his May 24 letter to the editor, Young stressed that the committee had no authority to enter into such an agreement without first conferring with the town meeting, and that the committee had done "more negotiating than investigation."

"I wish that time had permitted more study, but a lawsuit, already several times delayed, faces the town next Thursday," he wrote. "This is a suit brought against the town by the Bridgeport Hydraulic Company regarding the seizure of town roads."

Mr. Young warned that at the upcoming town meeting voters would have to choose "whether to settle or to fight," and that some members of the committee, himself included, had repeatedly opposed the idea of a settlement.

"A demonstration of courage would tend to bring the Bridgeport Hydraulic Company to a more reasonable attitude towards our town. The Bridgeport Hydraulic Company had things pretty much their own way for years, and they were seemingly going to do a rough shod over our town as they had over individual property owners. Little by little, however, they have shown that they realize that the town is really

Fig. 26. Copy of a torn poster, "Don't Take a Licking from the Bridgeport Hydraulic Company," from the Saugatuck Valley Association. Courtesy Weston Historical Society, Inc.

aroused, and that we possess fighting strength that the Bridgeport Hydraulic Company had not anticipated."

Weston's fighting spirit, Mr. Young continued, made Bridgeport Hydraulic company form a new attitude toward the town. "Having confidently started out by only offering to give us new roads for old, they then, with more opposition from the [Saugatuck Valley] Association, and some members of the Town Committee, upped their former offer by dangling some cash in front of the committee. This failed to dazzle us, so the Bridgeport Hydraulic Company once more raised their offer, this time to a glittering $37,500."

He called the $37,500 offer an insult to the town, suggesting that $100,000 "would be somewhat more in line with the damage that a dam and reservoir would do to this grand, unspoiled town of ours."

"If Weston still wishes to sell out to this group of shrewd business men who operated under a charter which ought to be stuffed and placed in a museum of freaks and quaint relics such as the dodo, the dinosaur and the bustle, if Weston sells out, why not get some real money instead of an insignificant $37,500."

Mr. Young's letter was as much a tribute to his motivational advertising skills as his devotion to the Saugatuck Valley Association. Nearly six hundred people turned out for the town meeting; it became so overcrowded that it had to be moved from the town hall to the Horace C. Hurlbutt, Jr., School a half-mile away. A May 27, *Westporter-Herald* front-page article described the meeting as tumultuous. Screaming matches ensued between those favoring and opposed to the settlement.

"Weston's Town Meeting Wednesday night was a classic, even in the long list of fighting town meetings. There were cops and constables, a near riot, community singing, violent disputes on the fringes of the crowd."

The division over whether or not to accept the agreement extended all the way to the board of selectmen. First Selectman Chester Coley offered the committee's majority report to accept the agreement. It was followed by a fighting minority report from Third Selectman Daniel R. Harvey, who accused Bridgeport Hydraulic of rushing the meeting.

"This company has throttled the entire development of the Valley Forge district," Mr. Harvey said. "For years they have held their threats over the town."

Harvey had long been an outspoken critic of the water company. His passion on the Valley Forge issue had not even been tempered by personal tragedy months earlier. His forty-three-year-old wife died after she reportedly fell asleep at the wheel and smashed into a tree on Weston's Fanton Hill Road one Saturday evening (*Westporter-Herald*, October 19, 1937).

"You will be selling out prematurely if you let them get away with this offer," he said at the town meeting."What do they offer? A road? We have a road, a perfectly good road we built and like to use. They offer a new road which will open up more of their property if they don't build the dam. They offer us $37,500. This litigation and delay will cost them more than that."

Mr. Harvey urged the townspeople not to believe the dam was a done deal in spite of the fact the Bridgeport Hydraulic Company had begun cutting down some trees in Valley Forge, and had begun construction on a road and a bridge. "They have just been staging a nice act over in Valley Forge," he said. "Don't worry your heads over that."

Cheers, shouts of approval and floor stomping from the audience greeted Mr. Harvey's comments, and "frequently waves of laughter swept the audience at some of his jibes."

"It has been years since so many people attended a town meeting in Weston," the May 27 article noted. "Many supporters of the Saugatuck Valley Association, which has been fighting the construction of the dam, came out from New York City especially to vote against the Hydraulic's offer. For more than a week a concerted campaign has been waged in the town by peole who wanted to accept the offer and those who did not. As much electioneering has been done, one old-time resident said Wednesday night, as any national or state campaign would have brought about."

Ultimately, the town meeting turned down the water company's offer of $37,500 for new roads but approved accepting $5,000 from the Saugatuck Valley Association toward court costs. The Saugatuck Valley Association celebrated with a party afterward.

Following Weston's lead, the town of Redding on June 6 rejected a substantial Bridgeport Hydraulic Company offer of $65,000 for the condemnation of certain roads. The town meeting's vote was 256 to 115 against the offer. As in Weston, the meeting had one of the largest attendances in the town memory. "Only legal voters were allowed in, and many of them had to stand," the June 7, 1938, *Westporter-Herald* reports. "Six state troopers were on hand." In addition to turning down the water company's offer, Redding, like Weston, approved a $5,000 appropriation for legal expenses.

At the meeting, Col. Edward Steichen, a photographer, is reported to have given one of the most stirring speeches. "They want to bury one of the most beautiful spots in America," he said. "It took mother earth 10,000 years to make the glen. They offer us $65,000 for it. Fiddlesticks!"

The Town of Westport, too, was now poised to officially enter the controversy. Perhaps it was because of Mr. Harvey's points made on May 27 that still echoed along the river banks from Weston to Westport—concern about the impact of the dam on the Saugatuck River.

"They offer us 1 million gallons of water over the dam every day," Mr. Harvey had stated. "My engineering friends assure me that that will amount to a trickle 18 inches wide and three inches deep, running as fast as a man can walk."

By mid-June a Westport public hearing was held at Bedford Junior High School, where residents listened to reports that the daily flow of one million gallons "is about one-eleventh of the amount of flow in the driest season of the year," according to the U.S. Geological Survey made off Ford Road over a five-year period (*Westporter-Herald*, June 10, 1938). A week later, hundreds of Westporters filled the junior high school auditorium for a special town meeting that officially brought the town into the battle. The town passed a resolution demanding eleven million gallons per day.

Perhaps the request for eleven times the amount was Westport's way of negotiating to receive at least half or one-third that amount. If that were the haggling logic behind the resolution, it worked. Ultimately, the water company agreed to a daily flow of four million gallons, which, while it would not guarantee the river all knew and enjoyed, would at least be better than a trickle far worse than a period of drought.

But for Weston residents and members of the Saugatuck Valley Association, quibbling over the flow from the dam was a moot point, especially if the dam were denied in court. Undaunted, many still believed they would prevail, and they had the big guns to prove it, including Governor Wilbur Cross who supported them. They refused to accept defeat. Months earlier John Orr Young, having launched a suit challenging the constitutionality of the Bridgeport Hydraulic Company's charter in the Supreme Court of Errors in Bridgeport, put it this way:

"We have never looked upon our previous setbacks in courts as defeats but rather as stepping stones to an appeal to a higher court where the real issues can be tried. This is our turning point in the success of our challenge to a public utility which has always nurtured the belief that it is invincible. We don't believe it" (*Westporter-Herald*, March 4, 1938).

They remained steadfast, and they remained strong, and so did Samuel P. Senior who, because of the power he wielded, proved Mr. Young and his association wrong.

9

The Defeat

Living in a time when environmental groups command great influence, spurring media attention that leads to legislation, it's hard to imagine how the Valley Forge battle was lost, especially with the Saugatuck Valley Association's long membership roster of movers and shakers. From Lillian Wald to John Orr Young to Connecticut Governor Wilbur Cross to the scores of writers and artists of worldwide acclaim, it seems that few individuals of note wanted the reservoir built and the storied valley lost.

Why then its watery fate? Is it that money talks, and ruthless big businesses always get their way? Is it that despite our belief in democracy, the judicial branch of government, namely the courts, wield far too much power? Was it a combination of both? Or was it that the judges knew something that others clouded by emotion could not—that Samuel P. Senior was an extraordinary visionary, a genius and a seer who could divine water needs decades into the future?

To examine what happened, one must look at how the Bridgeport Hydraulic Company went about its goal to set Mr. Senior's vision in stone, literally, in the form of a multi-ton concrete dam. From the time the water company first eyed the valley to its flooding, Bridgeport Hydraulic, as John Orr Young complained, acted as if it were invincible. One would think Connecticut's Governor Wilbur Cross would hold some power, if not influence, over just one of his state's many utilities. Ten women from Westport, Weston, and Redding thought so, and in early May of 1938 they traveled to Hartford to make a plea to Governor Cross to halt cutting and further condemnation. The women (never identified by their first names) included: Mrs. Everett Boughton, Mrs. John V.N. Dorr, Mrs. L. H. Goldhorn, Mrs. John Maurice Clark, Mrs. Edward Robotham, Mrs. William T. Wisner II, Mrs. W. T. Nichols, Mrs. Lyda Sokoloff, Mrs. William Gillies, and Mrs. Brinckerhoff (*Westporter-Herald,* May 13, 1938). In the governor's office, their official statement charged the water company with "ruthless abuse of its powers and disregard of the attitude of the residents." They also begged Governor Cross for "sympathy, aid and counsel," and invited him to tour the valley "and to witness the devastation of the large, ancient trees that had already been cut down."

"We have been held up by legal and red tape," Mrs. Dorr told the governor. "We beg aid. If a special session of the legislature cannot be called to halt this thing, we want your sympathy and moral support."

The women surely knew their last request would be granted, even if sympathy was all Governor Cross could offer. Years before assuming the state's top post Governor Cross had lived one year in Westport during a brief stint as principal of Staples High School, which served Westport and Weston students. Weekends, Principal Cross could be spotted hiking through Valley Forge and Redding Glen.

"Weather permitting, I will be glad to come down Sunday," he was quoted in the May 13, front-page story. "I have known the beauty of the Valley and Redding Glen for many years, and I regret that it faces destruction."

When Sunday came, the governor proved he was no fair-weather friend. Despite a rainstorm he spent the day traveling twenty-five miles back and forth through the valley from Weston to Redding. He gave speeches at both the Mark Twain Library in Redding and Weston's Dorr Mill, where receptions were held in his honor (*Westporter-Herald,* May 17, 1938).

"With a straight forward expression that the Saugatuck Valley can be saved from the proposed flooding by the Bridgeport Hydraulic Company, Gov. Wilbur L. Cross gave the fight being made for the Saugatuck

Valley Association his blessing and encouragement Sunday afternoon," the article ran. "During the tour he had passed groups of people standing on their lawns bearing placards, 'Help save our homes. The Bridgeport Hydraulic charter is unconstitutional. Help save our Saugatuck.'"

A large group of Redding residents greeted Governor Cross at the Mark Twain Library, where school children presented him with a bouquet of valley wild flowers. "My dear friends and children," he told them. "I came down here today to see the Saugatuck Valley. When I was a young strapling, I lived in Westport one year, and I had the most enjoyable stay. I was principal of the Staples High School. I came up to the valley often in those days, but that was a long time ago. All of you should be very happy to live here, provided you don't lose the valley."

The governor vowed to seek legal advise about any direct aid he could give the preservation effort adding, "But one thing I can say. Some restrictions should be placed on the private companies who operate under the legal opportunities given them under the right of eminent domain."

But even the governor's moral support was no match for the water company's much criticized, eighty-year-old charter that was reaffirmed court case after court case despite noble efforts and righteous speeches. Land acquisitions and condemnations were already a fait accompli. Trees were being cut. New roads were being blazed to replace those that would be washed away. The old colonial homes and mills would be burned to the ground within a year. And blasting would begin on the ledge near the southern-most point where the dam would be erected.

The Saugatuck Valley Association had lost. It wasn't a long, drawn out defeat. It was more like hitting a dead end on one of the valley's discontinued roads. After years of fighting and amassing legions of supporters, the war was simply unwinnable. Even for the celebrity crowd the fight had lost its luster and was no longer front-page news in the *Westporter-Herald*. From time to time undaunted association members would rattle some cages by calling for legislation to change the water company's charter. Without that change, they said, even residents' backyard ponds and swimming holes were potential targets of eminent domain. But exhausted from the long fight, residents' accepted defeat and turned their attention to a more pressing, global concern—

Europe on the eve of World War II. Full-page Saugatuck Valley Association advertisements had been replaced by full-page ads calling for war to stop the Nazis.

A brief town notice appearing on page 5 in the May 3, 1939, *Westporter-Herald,* "Valley Forge School District Is Formally Abolished" conveys the town's feeling of resignation to the water company.

"At a short and harmonious special town meeting held on Thursday evening in the Town Hall, voters of the town formally abolished the Valley Forge School District, which is being taken over by the Bridgeport Hydraulic Company for the projected Saugatuck Valley reservoir," it read. "The action, which marks the town's formal abandonment of the district, was taken upon a motion presented by First Selectman Chester G. Coley."

For Florence Banks, who ran the Valley Forge School, it might have been the painful end of an era. The school had been her only life since her fiancé had died on a European battlefield during World War I. Her only focus had been the children. But ever the educator, Miss Banks must have seen the Valley Forge School District closing as a form of commencement, too. From her small Cape house where she lived alone on Norfield Road, she drove ten miles each day to Norwalk, an industrial city on the Sound where she found work at the Roosevelt Elementary School. She later became principal, serving in that post until her retirement in 1955. Afterward, she remained an active member of the Weston Grange until she died in the late 1970s.

"She was very personable," recalled Peg McCullough, a Weston resident who also taught at the Norwalk school. "She was a neat lady, she was very good with children, and she was like an elderly grandmother" (June 12, 2004, interview). Despite her new life as a Norwalk educator, Miss Banks remains immortalized by the photo of the lone school mistress standing beside her children outside the Valley Forge School.

That frozen, tranquil image, however, like the peaceful pictures captured by Mr. Whitlock on his glass plate positives, faded quickly throughout Valley Forge by the Summer of 1939. The windswept mornings once redolent with birds singing and the river's rush had been usurped by grunting bulldozers, singing bandsaws, hacking axes, cracking timber, and continual blasting. The latter proved most traumatic for wildlife. Animals ran, flew, and slithered away.

"Natives of that vicinity report an influx of snakes this year" stated the front-page story, "Woman Bitten by Copperhead," in the August 18, 1939, *Westporter-Herald*. "The oddity had been explained by a reptile authority who said yesterday that that peculiarity is frequently noted where blasting has been prevalent.

"The snakes, he revealed, leave the immediate vicinity of the blasting," the story continued. "This situation might account for the appearance of the deadly reptiles here, he said. Contractors have been blasting several months in the vicinity of the proposed Saugatuck reservoir dam."

The story described how Grace Robinson, a "noted New York newspaper woman," was bitten on the ankle as she was walking a quarter-mile from her home one evening near Godfrey Street and Newtown Turnpike. Near collapse, she was taken to Norwalk Hospital where she was reported in stable condition.

An exodus of poisonous snakes and the threat they posed, however, was of no consequence to the water company determined to stay on Mr. Senior's timetable, nor to the workers who would clear the valley, help build the dam, and eventually build the tunnel to Easton. Hungry for work, these Depression-era forty-niners saw the valley paved with gold, as Bridgeport Hydraulic was paying a premium forty-five cents an hour.

10

The Dammed

The old farmhouse on Everett Road in Easton doesn't look much different than it did more than seventy years ago when eleven-year-old Olive would sit on the porch and hope to see an occasional automobile wind its way down the dirt road. She would keep a log of the makes of the cars, be it a Chevy, Ford, or Buick. Her father, a farmer from Sherman, Connecticut, had moved the family to the southern part of the state to be close to Bridgeport where he found work as a car salesman. Today, in her eighties and using a wheelchair, Olive lives in the same house with her husband, Bill Franklin. They moved back from Weston's Valley Forge in the early 1940s after Bill's three-year job there ended. He was one of the men who cut the trees, built the dam, and dug the tunnel.

"Most of the guys that worked on that job are gone now," said Bill, a towering yet soft-spoken man, during an interview at his home in June 2004. "At one time I knew all their names, but now that's gone. My memory's not worth peanuts now."

But with prodding from Olive, Bill's memories, scattered fragments at first, take form. His recollections run like water over a full dam, furiously at first before dissolving into trickles. Born and raised in Bridgeport, Bill Franklin landed his first job in the valley in the late 1930s at the Stowe Lumber Mill. There, he and and a dozen men swung axes

Fig. 27. Saugatuck Valley at the time of the dam construction, from the air. Courtesy Weston Historical Society, Inc., Herb Day Collection.

and pushed and pulled two-man saws to help clearcut the hillsides near the dam site. "There were no chainsaws then," he said. Their foreman was Irving Ruman, who each morning drove a pickup filled with axes, saws, and water to the site. All day the workers chopped and sawed, shouted "timber," and stripped the trees of branches before they rolled them down the hill toward the mill.

Bill recalled how one day a mill worker fed a log too quickly into the saw blade. "The log flew up and hit this guy and killed him. I remember I was down at Ruman's house, and they drove by me with the guy in the back of the pickup, but he was done. He looked pretty bad."

The man's name now escapes him, maybe because the death slowed the project by barely an afternoon. Though shocked by a grim reality, there was enough work sawing and chopping to keep workers' minds occupied and perhaps avenge their fallen comrade by taking swings at the trees. Once the hillside was cleared Bridgeport Hydraulic brought in a steam shovel to extract the roots. Otherwise it wouldn't be long before saplings began sprouting around the sawed-to-the-ground stumps. After the obstinate, probing roots were extricated, the soil-packed

Fig. 28. The Saugatuck Reservoir dam under construction. Courtesy Weston Historical Society, Inc., Herb Day collection.

clumps were left for the workers to hack into kindling. It was an arduous task, especially when the temperature dropped. One March morning Bill Franklin faced a patch of roots that were frozen solid. They were impenetrable to ordinary axes. "We had to use pickaxes on them and then we tried to burn what we could to keep warm, but it was green and didn't burn too well."

The hillsides at the dam site were just the beginning of what would amount to a more than eight-mile clearcut.

"There were a few hills in the valley that are the islands now. They were so thick, you could get lost in them. The water company brought in a crew from Maine, and they were expected to clear those hills in no time. But they were used to cutting pulpwood, not hardwoods, and this took time."

For his backbreaking labors, Bill earned a good salary in 1939, thirty dollars a week. Soon after, he married his longtime sweetheart, Olive, who had been living on her parents' farm in Easton. The newlyweds moved to the valley and into a small two-bedroom white house on Anson Morton's property just south of the dam site. The twelve dollar monthly rent, says Olive, was very reasonable, even for the Depression.

"It was the most wonderful place in the summer. There were three other couples in Anson's houses, and we women would all socialize. In the summer we'd spend days at the beach in the river and wait for the men to finish work. They'd come down after work, take a bar of soap wash up right in the river."

Olive also enjoyed fishing the Saugatuck, especially for eels. Sometimes she would visit the home of Clinton Hull, Clinny to his friends, who had huge eel nets set in the river. From the banks of Hull's property and with his permission, Olive would often cast a line to bring home dinner.

Even though Bill and Olive's livelihood came from the valley's ultimate demise, the place had begun to grow on them. Sure, they lived in one of the Morton's houses south of the flood area and immune to eminent domain, but the valley location is what made it special. "Oh, God, it was wonderful," said Bill, "There were two nice streams that came into the river. It was all green and lush and wild and nice with old houses."

Local color could also be seen in the characters in and around the valley. One was Anson's mother, known only as "Mrs. Morton," who was cut from a frontier cloth so self-reliant she refused any shows of chivalry around the property. Heaven help any of the male tenants if they offered to cut her cord wood for her stove for cooking. For that task she had a Bradley ax that suited her just fine. Another character was a water company supervisor known as Mr. Merrill. While the men were busy chopping, Bill recalled, Merrill had a habit of visiting the ledge where the men kept their lunchboxes. He would sample their lunches as if he were at a buffet. When the men broke for lunch, they would find half-eaten sandwiches and others with bites taken out. "One guy got back at him one day by filling a sandwich with hot peppers. Merrill ran out and down the hill looking for water, and he never touched anyone's sandwich after that."

Then there was another fellow named Garrity, Bill remembered, who liked to drink, not just after work with the men at a Redding pub owned by a guy named Rainie, but on the job. Some days Bill would find Garrity passed out under a tree, near the dam, even in the tunnel during construction. Finally, there was a man Bill Franklin remembers only as Davis. He was a valley subsistence farmer who lived in a shanty. The only enterprise he showed outside his small patch of land was to come to the Mortons' property from time to time for odd jobs when he needed money.

Living and working with these people, the Franklins developed a strong sense of community and a love of place, just like the early settlers did. That's when the water company's ultimate objective began to hit home.

"One day when the ladies and I went to the beach, we saw that Bridgeport Hydraulic had filled the river with these huge trees they had cut so we couldn't swim in it. It was a dirty, rotten trick."

Olive's only consolation was that her husband's crew had not cleared that area. It was another crew logging the northern part of the valley. Not that Bill's ax would lie motionless if Ruman wanted his and Olive's cherished area cleared. A thirty-dollar-a-week salary, sometimes forty dollars with overtime, helped salve any strained emotions. To the Depression's downtrodden only one kind of green really mattered. Bill later found other work near the dam site. While imported lumberjacks cleared the northern part of the valley, he landed work at the biggest job of all, the dam.

The first phase, however, involved more grunt work. He and others had to shovel pulverized ledge and dirt tossed up by the dynamite into dump trucks. The job was a lot dirtier and more strenuous than just felling trees. Bill also encountered a new menace due to the blasting, a preponderance of copperheads. "I wore boots up to here," he said pointing to his knee. "They wouldn't go looking for you, but sometimes it was hard to avoid them."

When that job ended, Bridgeport Hydraulic brought in another expert, Bill's new boss, Bob Crowe. "He was in charge of the whole thing." Bill described Crowe as a tobacco-chewing fat cat who toured the area in his tan coupe. "He'd always be spitting tobacco juice, even from his car, and he didn't care if it got all over his door." Crowe's duty was to oversee the construction of two, more than 200-foot high, 30-foot-wide steel-girder towers on both sides of the river. Workers strung steel cables on a pulley between the towers. An electric motor worked the pulleys so the metal cables would travel back and forth, like those on a ski lift. Suspended from these cables were massive metal tanks. Bill Franklin suspects the vats held more than two hundred pounds of freshly mixed concrete hoisted up from the ground after it

Fig. 29a., 29b., 29c., and 29d. A cement vat is hoisted on metal cables above the dam to where it will pour the dam's next tier. Courtesy Weston Historical Society, Inc., Herb Day collection.

Fig. 30a. and 30b. Workers spread the newly poured cement and later work air-powered vibrators to excise air bubbles that would compromise the dam's structural integrity. Courtesy Weston Historical Society, Inc., Herb Day collection.

had been properly mixed with river water. The pulleys would propel the vats over different areas of the wooden forms to be filled. Then Bill or another man would pull the vat's hand cord for the concrete to pour into the forms, which would stretch the entire length of the dam, seven hundred feet. At its base the dam was one hundred feet wide, but the width would taper to twenty feet when it reached the top.

"Sometimes we poured only a foot a day," Bill recalls. During these initial stages the river had been diverted around the site. "As we poured that bottom, they built an opening into it, a gate, an eight-foot-wide valve."

Bill's task was to run an air-powered vibrator over the soft cement. Comparable to a jackhammer, the vibrator had a wide metal plate at its base. He would aim the rapidly pounding plate at the unset concrete to free any trapped bubbles lest they leave air pockets to weaken the dam's structural integrity.

Fig. 31. Another view of the dam under construction. Courtesy Weston Historical Society, Inc., Herb Day collection.

"One day when I was working the vibrator on the west side, I heard all these guys hollering. I looked up and there was the cement bucket swinging right at me. I ducked, and it just missed my head by an inch. That happened because the signal man had given the wrong place for it to stop."

As the dam rose, the clearing continued, and the river was slowly diverted back to its original course. The valve gate at the dam's bottom was regulated so the rising water would not reach the high point where the new cement would be poured.

Bill had assumed that all Valley Forge residents had pulled up stakes during this period. But Mrs. Pierson, a widow who lived alone, refused to leave. It was her home, and she was willing to go down with it despite repeated attempts by the water company to evict her.

"The water was coming up, it was about seventy-five feet from her house when Bridgeport Hydraulic finally got the courts to let them remove her," says Bill.

All of her furniture was put into storage, and the state agreed to pay Mrs. Pierson's rent at the home of a valley resident on high ground. Weston resident Ernie Albin, who as of this writing lives in an old

The Dammed 75

home on the corner of Valley Forge Road and Newtown Turnpike, believes Mrs. Pierson might have been the mysterious old lady always wearing a long black dress. He remembers the strange woman who in the early 1940s rented space from the late Minerva Morehouse on Newtown Turnpike. The woman always shied away from people and would immediately disappear when company came. "When we would go to visit the house next door, she would always leave, descending the steps in this long black dress."

Another stubborn Yankee dead set against leaving his property was old Mr. Davis. Shanty or not, it was his home. But after repeated attempts to evict him, Davis acquiesced, willing to give life in a Westport apartment, all paid for by the water company, a try.

"He was there only one night, and he was gone," Bill Franklin said with melancholy disbelief. "He never had a gas stove. He didn't know what the hell it was and probably didn't know the gas was on when it wasn't lit. That's what killed him."

Mary Ann Barr, a director at the Weston Historical Society and a lifelong town resident, says there's a folktale in town that at midnight you can see a foggy, ghostlike figure crossing the small cement bridge just below the dam and heading to the Morton property. "Maybe it's Davis," she says.

Weston resident Ted Coley was just twelve in 1940 when the dam was completed and work on the tunnel had begun. Though a long time ago, just mention the words Valley Forge, and Mr. Coley is besieged by a paroxysm of anger and hopelessness as if he were the victim of a heinous, violent crime.

"You see, as I get older it gets worse for me," he said. "It's this whole idea of eminent domain."

His grandparents, George and Matilda Hicks, had a home on seven acres in the flood zone. "It was an old-fashioned farmhouse with a barn in the back," he recalled. "They had grape arbors, a hand-dug well with a bucket and some cranberry bog."

"I remember when they set the barn on fire," he continued. "My mother, her name was Anna, was so upset."

It was a loss that haunted Ted Coley's grandparents and his mother until their deaths. Before the water company burned down the Hicks house, Anna Coley paid to have the well stones removed by steam

Fig. 32. Raging waters near the dam's spillway. Courtesy Weston Historical Society, Inc., Herb Day collection.

shovel and carted to her own home on Lyons Plains Road. Ted Coley, who now lives in the house, is reminded of the loss every time he steps out his front door and onto his well stone patio entry. Not much else from his grandparents' house was saved.

"The trouble is they stole a lot of stuff," he said of the workers brought in to take down the houses.

Olive Franklin shared the same thought that water company employees picked over the houses, "getting all the good stuff," including chandeliers, doors, mirrors, even furniture. While outraged at the looting and saying, "I hate the water company," Ted Coley holds equal contempt for the valley residents who sold out in the early years.

"Senior started out by getting flunkies to take our land from us. That is why it has always been a bitter thing for me. Then they had some sneaky natives. Anytime they needed money, they'd sell some land. They'd start by selling their wood lots they didn't think they'd need."

One of the so-called sneaky natives was Charles Rowland Morehouse, a shill for Bridgeport Hydraulic who, after he sold his land, amassed substantial water company stock by touring the valley like a native son convincing others to sell.

Fig. 33. Workers digging the tunnel from the Saugatuck Reservoir through Pop's Mountain to Easton's Aspetuck Reservoir. Courtesy Weston Historical Society, Inc., Herb Day collection.

"He would go down and talk to them and say, 'This looks like a pretty good deal,'" his great-grandson Ernie Albin told me in June 2004. "I wonder why he wasn't lynched."

Ernie described Charles Morehouse as a farmer by occupation, who in reality was a wheeler-dealer in property. Over the years Morehouse had purchased and resold more than two-thousand acres in Weston and Redding. Mr. Morehouse also didn't lose the family homestead, which was out of the flood area and beyond the reach of the long arm of eminent domain.

While the valley sat within Anson Morton's backyard, he, too, did not have to fear losing his family's houses, which were below the dam. In fact, between rents from workers and regular jobs as an electrician for the project, Anson Morton found the building of the reservoir a financial boon. His big break came when Bridgeport Hydraulic began to drill the tunnel through Pop's Mountain and then Flirt Hill to the Aspetuck Reservoir in Easton. As the tunnel progressed, Anson was in charge of the lighting and wiring the tracks for the electric motor–powered trains that moved the drills and blasting equipment and carted the gravel out. Bill Franklin also found work digging the tunnel, his most lucrative job.

"I was averaging about forty dollars a week, and with the tunnel I got a raise, and we got bonuses for digging so many feet. Some weeks I got as much as a hundred dollars."

The tunnel started on the eastern side of the dam near the drain valve and ended up in the Aspetuck Reservoir in Easton. From start to finish the tunnel would run two-and-a-half miles. The hardest part, Bill Franklin said, was the mile stretch through Pop's Mountain. They drilled, shoveled, and swung pickaxes in the eight-foot diameter hole that was mostly ledge. "Sometimes we made two hundred feet a week. We ran into trouble when we hit flintstone, and they had to haul in a specialist with explosives."

They also ran into trouble when they drilled underneath wells in the mountain. Sometimes they would leak, and the holes would have to be patched with a special type of sealant that would harden in water.

The engineering specifications of the tunnel called for it to rise one foot every mile. The idea was that when the water level reached a certain level in the Saugatuck Reservoir, the water would be forced down into the tunnel to travel the distance to Aspetuck. It enters Aspetuck by way of an aerator that takes the form of a high spraying geyser still visible off Route 58 in Easton.

One would assume that the tunnel would house a pipe to transport the water, but, says Bill Franklin, "the tunnel was the pipe," and for that purpose it had to be lined with concrete. "We would start on the bottom and put it on halfway around to a wooden form where there would be a ledge, and when that set we did the top part."

By 1941, when the tunnel was finally completed, Bill and Olive packed up, bade farewell to what was left of the valley and its rising waters, and moved into her parents' house in Easton. Bill had a lot of money saved, and thanks to Anson Morton landed a new job at United Illuminating. The pay, forty dollars a week, wasn't as good as it was working the tunnel, but it was a steady income.

Looking back, he displays a modest, quiet pride at his versatility during the entire project. The only job he had refused was the offer to help exhume bodies from the Burr Cemetery in the northern flood zone.

"I wouldn't touch it," he says "Others were happy to have the work, and happy about what they might find digging." Some of the graves were more than two hundred years old, "the boxes were all gone, and maybe they'd find the bone of an arm."

Any of the grave diggers hoping to find jewelry and other gold probably had their treasure hunting dreams quashed by two of the men overseeing the project, Ted Coley's grandfather George Ricks, and Ted's uncle, Sam Ricks. They only accepted the job to make sure the village's dead would be respected and could rest in peace, if not in the valley. The remains, Ted remembers, were all placed in small boxes and reburied in Redding's Hull Cemetery, where the graves and tombstones were lined up in rows in military fashion.

Bill and Olive Franklin moved to Easton before the entire valley was flooded, and they had vowed not to look back. But within a year's time their curiosity won out, and one Saturday afternoon they decided to revisit the area. "In 1942 I drove around it on the new road, and I couldn't believe it."

There was also something about the reservoir even Bill couldn't comprehend.

"It looked like it had always been there."

11

At Peace at the Reservoir

For Joe Haines, the Saugatuck Reservoir might as well have been in existence since the beginning of time. Born in 1937, the only recollections this Easton resident has are of the reservoir's expansive water surface, its craggy, tree-lined shores, and the woods that surround it.

"I don't think there's more than fifty square feet that I haven't walked over," Mr. Haines, a barrel-chested, Yankee outdoorsman boasts about the Saugatuck Reservoir and the contiguous Trout Brook Valley.

Joe Haines spent the past thirty eight years as a game warden–patrol officer for Bridgeport Hydraulic, and he walked the woods daily until he retired in 2002. Licensed by the state to make arrests and equipped with a sidearm, Joe's job was to make sure the region's water supply remained protected from all who would do it harm. In today's culture that would translate into protecting a water resource from terrorism. But in less worrisome times and for nearly all of Joe's career the focus was on the occasional fish and game poacher, those bold enough to steal a swim or launch a boat, or worse, dump garbage. Joe even worked with local police to help track burglars who fled into the woods thinking law enforcement could never find them in the dark woods. What these fugitives underestimated was Joe's rooted connection to the land, his grasp of all the subtle nuances of the forest, his emotional, spiritual relationship to nature typically associated with Native American guides. "I could never survive in the city," Joe says, "but I'm right at home in the woods."

Over the years, he has patrolled water company lands in all of Fairfield County, as well as in Litchfield County as far north as the town of Canaan near the Massachusetts border. But of all the lands he has traveled, Joe Haines holds the greatest affection for the acreage surrounding the Easton and Saugatuck reservoirs. Perhaps it's because it is in his backyard, or that it *is* his backyard. Few, if any, can stare at a forest and say they have met each tree and watched them grow up. Joe Haines can.

So don't expect Joe to wax sentimentally about Valley Forge and the entire Saugatuck Valley before the flood. In his near forty-year trek through these woods, his only connection with Valley Forge has been archaeological. He unearthed blue insulator caps next to the stumps of telegraph poles that had stretched down into the valley from Newtown Turnpike on the Redding border. In the same area he came across cracked, leaning, and half-buried eighteenth-century headstones from the part of Burr Cemetery above the flood line where remains had not been exhumed. On hills near Pop's Mountain he would find charcoal pits, holes dug to manufacture the fuel for forges. He also gleaned images of the valley from conversations with an old-timer on his route, Anson Morton.

"The woods are full of old charcoal pits up on the hill," says Joe. "Anson told me he and his father would spend all winter cutting wood to make charcoal."

The practice had been going on in the valley for two centuries. "They dug a large hole, filled it with wood, and when the wood was burning, they would throw in dirt to smother the fire just enough. This starves off the oxygen, and it would turn the pieces of wood to charcoal. There was no importation of coal, so this was the next best thing."

On the upper end of the reservoir Joe discovered a cement slab foundation of what had been a cow barn. "You can still see some of the stanchions sticking up at low tide. These were metal pipes curved into an oval. The cows heads would go into it, and in front of it was a trough with feed. And when the water goes down real low by the islands, there are five of them, three or four house foundations come up.

I found a lot of old bottles out there." About two thousand feet inland from the Redding shore is a rusty, timeworn steam boiler that had been used to run a sawmill. "It's just sitting there on the ground."

Then there are the remains of roads that resemble mere driveways to the water's edge. The most prominent one is near the dam and enclosed by a chainlink fence with the sign, "Angler's Dock." It looks like a boat launch for the water company officials, though in fact it is what remains of Valley Forge Road. An underwater amphibian vehicle could conceivably drive down Angler's Dock and continue into the valley and its submerged roads.

There are times, especially when Joe recalls the conversations he had with Anson Morton, that he feels badly about the valley's demise. "The hardest thing they did was to condemn the property."

"But the reservoir wound up being used, and it's very important. The Hemlocks Reservoir is not that deep, and without the Saugatuck supplying the Aspetuck to back it up, it would be sucked down dry. For the few people that did lose their homes, you have to think of the hundreds of thousands of people over the years who were satisfied with the drinking water.

"I would say it did help the environment," Joe continues. "There's a lot of wildlife it serves. There are wild turkey, deer, rabbits, plenty of squirrels, and an occasional bear wanders in. And the reservoir itself is a tremendous fish hatchery. You have trout, bass, sunnies, and walleye. Someone recently pulled out a walleye near the dam up to twenty-nine inches long. And if there was no reservoir, the valley and land would have been developed."

Generations of Easton and Weston residents, oblivious to the Valley Forge battles, have felt the same way. The water company's control of thousands of acres of pristine forest to assure clean water seemed a guarantee these lands would never fall to development. Many saw water company ownership as a boon to Connecticut's natural resources, a kind of quality-of-life insurance for a state already overstressed by population increases.

It wasn't long before Valley Forge was gone and forgotten. Many of the stories of the residents' struggles, desperation, and heartache at the hand of eminent domain had died with them. The few who remembered, like Ted Coley, were written off as curmudgeons. What mattered to the new generation was the peace of mind that all the untouchable land offered the surrounding communities. It was part of the water company's continuous emerald chain that stretched all the way to the Massachusetts border. Nothing would ever change that. Like the ocean always being there or the sun rising each morning, they could count on the land remaining undeveloped.

Yet in the early 1990s, that, too, would change.

12

Another Valley, Another Time

Good intentions sometimes go awry, and acts that were designed to help can wind up causing more difficulties. One such good intention was an amendment to the U.S. Clean Water Act that required reservoirs to be treated by complex filtration plants to rid the water supply of acidity caused by leaves and other all-natural contaminants. Nature, it seemed, was not a good enough caretaker, and technology would have to set things right. Ergo, many of the lands that Bridgeport Hydraulic seized from the farmers to safeguard the water supply, some reasoned, were not that great a shield after all. In fact, they were unnecessary with the rise of filtration plants and could be sold off.

In 1991 Aquarion, a publicly traded company, owned the BHC, and shareholders and officials saw dollar signs when they realized how much saleable acreage the company and its subsidiaries owned. The water company also had to pay for the expensive filtration plants mandated by the federal government. Each would cost between $40 to $50 million. BHC was already building one across from the dam at the Hemlocks Reservoir and at the base of the Easton Reservoir. To raise the money, Aquarion-owned BHC began carving up parcels to sell. One was a more than 700-acre tract just over the hill from the Saugatuck Reservoir. Known as Trout Brook Valley, this land had also been purchased in the early 1900s. Robert Kranyik and Verne Gay in their book, *Trout Brook Valley: Forever Yours,* write:

> These plants were very expensive to build, each in the neighborhood of $40 to $50 million. . . . Thus Trout Brook Valley, one of the largest single tracts owned by BHC, and controlled by Aquarion, was under consideration as a property to be offered for sale. Over 700 acres of nearly pristine land situated between the Saugatuck and Aspetuck reservoirs, though draining into neither, was now in jeopardy. (29)

Thus the stage was set for another water company initiative poised to alter life as it had evolved in reservoir communities. With it would come new battles just as heated as the ones for the lost and forgotten Saugatuck Valley.

The fact that the land would even be considered for sale was symptomatic of the boom times Fairfield County experienced in the mid–1990s, a trend that has continued through to this writing. The region's building and construction industry, in the doldrums since the crash of 1987, became reinvigorated with Wall Street's rebound and declining interest rates. Acreage that had gone unnoticed by passersby suddenly grabbed attention when the ground was initially scored to make way for construction. Within no time shells of massive homes appeared and more would follow wherever there was buildable land. And as demand for property increased, towns previously considered too far out for New York City commuters had come into their own. New fortunes, the relocation of New York City trading houses into suburban corporate parks, not to mention telecommuting, made land in Weston and Easton hot properties. In 1995 it was difficult to find a two-acre building lot in Weston for under $300,000. By 1998, two-acre parcels off Weston's Norfield Road by the center of town were selling for $750,000. As a result, developers viewed Easton as the next real estate frontier. Officials at the publicly traded water company knew that Trout Brook Valley, in Weston and Easton, was worth a fortune.

Jack McGregor, Aquarion's chief executive officer, had been exploring the idea of transforming the valley into a golf course and luxury home enclave since the early 1990s. But the terrain was too difficult for the water company to tame. Still eager to profit from the property, he then made an overture to sell it to open space advocates, who under Connecticut law have the right of first refusal when a utility sells its land.

Kranyik writes:

> In April 1994, a meeting was held at the Bluebird [Easton's only restaurant] between Bruce LePage of the Aspetuck Land Trust and Jack McGregor, CEO of Aquarion. Jack informed Bruce that Aquarion had a design for Trout Brook Valley which included homes on the ridges and a golf course in the valley. He went on to say that Aquarion had decided not to build the golf course due to the formidable terrain. Jack then offered to sell some 400 acres to the Aspetuck Land Trust. Bruce contacted Gail Bromer, Chair of the Easton Open Space Task Force to propose buying the 400 acres. Gail Bromer and Princie Falkenhagen, president of Citizens for Easton, discussed the matter and concluded that the entire valley would be purchased if possible. (34)

That June the matter went to a town public hearing. The town of Easton, however, expressed no interest in spending its own money to purchase open space at that time. Still, the Open Space Task Force explored different funding options, such as foundation and government grants. They proved unsuccessful, as did the efforts of the Aspetuck Land Trust, a private nonprofit organization aimed at preserving open space in Easton, Fairfield, Weston, and Westport.

As time progressed, and with no major development looming, the Trout Brook issue remained quiet. In fact, by the spring of 1995 interest in Trout Brook Valley was totally eclipsed by a planned Aquarion sale much closer to home, the highly visible 27.8 acres on the corners of Routes 58 and 136 known as Easton's Four Corners.

The Aspetuck Land Trust began an effort to make Easton residents aware of the possibility that the Four Corners might be sold. Since the land trust had a right of first refusal under Connecticut law, a fund-raising campaign to save the Four Corners was initiated. Fund raising moved ahead. There was considerable support and interest within Easton, including support from the Easton Historical Society and the Easton Garden Club. Bumper stickers were sold and displayed on many autos in town, and there were articles in the newspapers (Kranyik, p. 35).

Though Easton had put Trout Brook on the back burner while it was absorbed in saving the Four Corners, the water company had not. For months Aquarion had been in negotiations with National Fairways, Inc., a golf course management company, and by early 1997 they were close to an agreement. In February, Aquarion announced the pending sale of Trout Brook to National Fairways, Inc., of Greenwich for $14.2 million. The company's designs called for an 18-hole golf course and a gated community of 103 homes priced at more than $1 million.

The news stunned many of the region's open space advocates, as well as Easton residents. The character of the town would dramatically change. Easton, of course, would have to approve the development in accordance with the town's zoning regulations. To sweeten the deal Marc Bergschneider, National Fairway's president, made a public announcement that his company would purchase and donate the Four Corners property as open space to the town. This offer, however, Kranyik writes, "angered many Easton residents who saw it as a bribe to alleviate any opposition to the proposed development.

"During the summer of 1997, following the dramatic announcement by Aquarion, things were unexpectedly quiet. Most residents of Easton and Weston who knew about the project assumed that the golf course and housing development was a 'done deal.' But things were about to take an unexpected direction" (Kranyik, p. 36).

13

Casting About

That summer, as I stood on the shore of the Saugatuck Reservoir and fished, I stared up at the hills wondered how much longer this feeling of being in the Adirondacks, as my father noted, would last. The crests of the hills, I had heard, would be lined with behemoth houses. Every time I tried to envision them, I felt that with suburban sprawl there was little you could count on to stay the same.

In Easton, the old apple orchard on land leased from Aquarion was nearly decimated, most of the trees cut down. I can still envision walking with my wife and in-laws alongside the apple trees, my father-in-law studying and smelling each apple before putting it in the bag. I can still hear the grunting tractor and feel the jostling, dusty ride out to the pumpkin patch, my son, brother-in-law, nephew and I being bounced in that hay-lined wagon in tow, my wife, sister, and daughter by the shed where they weighed and paid for the apples. But time moves forward even for those of us who prefer being freeze-framed by sentiment. Now that pasture was gone, and it would most likely become a fairway I couldn't afford to set foot on.

This loss and the pending sale of Trout Brook was a story worth telling, and since I wrote feature stories regularly for the Sunday Connecticut section of the *New York Times,* I contacted my editor one

Monday morning in early July. Dick Madden, a former Connecticut reporter-turned-editor, wanted the story. He lived in another rural but changing Fairfield County town, a community he chose despite the long daily commute to Manhattan. He knew the state well, and like so many residents he could see the Connecticut countryside changing and not always for the better.

I should note that despite my strong personal views, I set out to write an objective, not an advocacy, reaction piece to the Trout Brook sale. I wanted to explore the ramifications from both sides. After preliminary phone calls, my first stop was at the Bluebird restaurant. It is a very countrified and casual place located next to the Bluebird Garage, one of the town's three filling stations. The Bluebird is divided into two areas, one with tables, and it has a separate entrance for the counter and takeout. I chose the counter. Everyone at the Bluebird seems to know each other, so cozying up to a cup of coffee and making conversation is easy, even for a newcomer, or in this case, a reporter.

The first person to talk about the development plan was Julia Pida behind the counter. As she refilled my coffee, Mrs. Pida took a straightforward, business-like approach to Mr. Bergschneider's plan.

"It will bring in business," she said, identifying herself as the Bluebird's owner. "And as far as development goes, a few years down the road it will happen anyway."

Nearby was Louise Kranyik, a waitress in Mrs. Pida's employ who had lived in Easton for more than thirty years. Her home was right near the proposed golf course.

"We just don't want the neighborhood ruined," she said. "Our road is a scenic road. We just don't want to have seven hundred cars going down the road."

I later found opinions divided throughout wooded and rural Easton where the water company owned approximately 7,000 acres or 42 percent of the land mass. Some, like Anthony Colonnese, then first selectman, lauded the idea as a boon to the town's tax base. Each million-dollar home would increase town coffers by $10,000 a year in property taxes, he said. Mr. Colonnese's support might also have been tempered by the fact that he was an avid golfer, though his town salary seemed to preclude his joining Mr. Bergschneider's club with a $100,000 fee.

True, the first selectman said he felt a tinge of sentiment about the development underway at the site of the old Aspetuck Valley Orchards. "That goes back as far as I can remember," he said. "My folks use to go down there to the apple barn to buy pies. But things change. Nothing stays the same."

But another town official, Gail Bromer, blanched at the proposed sale. A member of Easton's Conservation Commission and chairman of the town's open space committee, Ms. Bromer called Mr. Colonnese's enthusiasm about an increased tax base unfounded. She said more houses would mean an increased drain on town services, including the need for more police officers, possibly even a new school. In short, the development would be a wash as far as taxes went, while Easton would see its rural character changed forever.

"The thing people keep asking me is how can a water company sell their land since they acquired it to protect the water," Ms. Bromer said. "They've been given preferential treatment all along in terms of taxes, and now they're going to sell it at market prices." Bridgeport Hydraulic, she said, had "turned into a development company."

Someone who felt strongly about this was Curtis Johnson, an attorney for the Connecticut Fund for the Environment. Mr. Johnson could not reconcile the fact that property obtained by eminent domain, or the threat of it, in the name of the public good could be sold for company profit. Not only that, but the land would ultimately be used for a very nonpublic use, a gated community and a pricey, private club.

I asked Daniel Neaton, vice president of Aquarion's real estate division, about claims that the development was not for the public's benefit. He justified the company's actions by saying any profits from the land sales "will be reinvested in the utility company, and that will offset consumer rates." He went on to say the Trout Brook Valley property, comprising 685 acres in Easton and 45 in Weston, was part of 2,600 acres of off-watershed land the company was selling in Fairfield and Litchfield counties.

Such concerns, I soon discovered, were not confined to Fairfield County. Linda Cardini, director of the Connecticut Rural Development Council in Winsted, Connecticut, told me many towns were grappling with the same issue because throughout the state public and private water companies had owned 133,000 acres. Now, it seemed, water companies had gotten into the real estate business.

"Small towns like Easton thought they had lots of protected open space, and that's contributed to their rural character," said Ms. Cardini. "Now they're realizing some of this land is not protected indefinitely, and that water companies are within their rights to sell these Class III lands, possibly Class II, if they get permission from the state."

Paula Pendleton, an analyst for the Connecticut Department of Public Health, explained these land classifications. Class I lands were defined as those within 250 feet of the high-water mark of a reservoir or 100 feet of its tributaries. These lands would be protected at all costs. A Class II designation was for any property beyond the Class I lands but still within the drainage basin. Class III were the off-watershed lands that had been purchased along with the other lands and were not subdivided when the land was obtained, usually under or in the shadow of eminent domain. Such were the cases with Valley Forge and Trout Brook Valley.

But again, these huge tracts of land in light of the new federal Environmental Protection Agency laws requiring multimillion-dollar filtration plants, seemed superfluous, economically foolish to keep if they could be cashed out. "Water companies no longer feel they have to have extensive land holdings to act like a buffer," Ms. Cardini told me, "because they have to treat the water before it goes out to the pipe."

Just a few months earlier BHC had closed on the sale of 471 acres of Class III property in Shelton for $7 million. The city purchased the land as open space out of fear that it would be sold for housing developments that would stress town services and increase taxes. The $7 million appropriation and bond authorization had little trouble passing a citywide referendum.

But Easton, with no industry and a commercial base of just a few businesses, like the Bluebird, was unable to come close to raising the $14 million needed to purchase Trout Brook Valley.

Bruce LePage, executive director of the Aspetuck Land Trust, had months earlier suggested that Easton buy the land for $5 million in town bonding, state money, and about $1 million raised privately.

"We talked to each of the four towns in the area, and we quickly found out that there was no money available," he said. "And they [BHC] wanted $14 million, and while we were waiting for a miracle, Mr. Bergschneider stepped in. What I'm upset about is that the towns couldn't buy it themselves, and that there was no money available for the towns to come up with a regional plan."

While upset about the prospects, Bruce LePage conceded that a golf course was still better than high-density housing. First Selectman Colonnese felt the same way, saying a golf course was nonetheless considered open space under state statutes. Making the proposal more palatable for Mr. Colonnese was that Mr. Bergschneider had been an Easton resident for ten years. He vowed he would be sensitive to the town's character.

"First of all, this property is over two ridges, nobody sees it," Marc Bergschneider told me in a phone interview. "Secondly, we are conforming to zoning by building private, rural roads—no curbs, no lights—and we have a golf course as open space."

In addition to the planned 200-acre golf course, a 12,000-square-foot club house and 103 houses each set on three acres, Mr. Bergschneider said he would also set aside 200 acres of Trout Brook Valley as open space. He was also simultaneously in negotiations with Weston's First Selectman George Guidera to sell Weston forty-five acres for open space once he owned land.

"So this is really low density," Mr. Bergschneider stressed. "My basic argument is that we are conforming to the way Easton is."

He seemed confident about getting town zoning approval and said that construction would probably begin in the spring of 1998. Already, National Fairways had constructed fifteen highpriced, private golf courses nationally. One was the Hudson National Golf Club in Croton-on-Hudson, New York. When it had opened a few years earlier individual memberships were $100,000 but had since risen to $162,000, he said.

"We dynamite-blasted more than 100,000 cubic yards of rock during construction," Mr. Bergschneider would later be quoted in literature circulated by the Coalition to Preserve Trout Brook Valley.

Another trump card Mr. Bergschneider held was that his development plan had been gaining acceptance as the lesser of two evils. "This land is going to be sold and developed even if proponents defeat my proposal. The only question is to whom. Residents need to consider who's in line behind me" (Coalition).

In retrospect, his thinly veiled scare tactic was reminiscent of the sales pitch made to farmers sixty to seventy years earlier when Bridgeport Hydraulic officials urged Valley Forge residents to sell at lower prices before condemnation proceedings made the real estate values drop even further.

Meanwhile, Trout Brook Valley was already being advertised. "Introducing Trout Brook Valley," one ad read. "Some of the most dramatically beautiful land in Connecticut. Enclosed in a cool forest. Behind high bluffs. Spectacular views and open space. Set between the Saugatuck and Aspetuck reservoirs."

National Fairways was also looking for a pitchman, and for that they approached a promising young artist, James Prosek, author of the book *Trout: An Illustrated History*, a text that includes his watercolors of many different trout species.

I had first met James a couple of years earlier, when his book was in its planning stages and he was in search of a publisher. His sister Jennifer, a public relations executive, had casually hinted that her young brother, who had just finished his freshman year at Yale, might warrant a good story. I wasn't sure, but I agreed to meet him.

When I did, I was awestruck. He had journals chronicling nearly every river, lake, pond, and stream that he had fished in Connecticut, and what he had caught and where. Each fish had been logged according to subtle nuances in speckles and hues and then recaptured by James's brush. Having fished numerous streams and lakes throughout New England as well, James was a combination sportsman, fish biologist, and artist. I became even more intrigued as I watched him prowl an Easton stream, his eyes almost feral as he searched for bubbles and studied the types of larvae under stones in the stream. I watched him gingerly move upstream making careful, deliberate steps even in hipwaders. The line of his fly rod whipped back and forth faster than a snake's tongue. In a few minutes James Prosek had a trout on the line. He photographed it, and by the end of the afternoon the fish had been immortalized in a watercolor.

I remember calling up Dick Madden to tell him I had just spent the afternoon with "the Audubon of the fishing world." The moniker stuck as my story with the same headline ran on the front page below the fold of the *New York Times* Connecticut section. It wasn't long afterward that James found a publisher and went on to be respected na-

Fig. 34. The planned 12th Green in Trout Brook Valley. Photo by Paul Avgerinos.

tionally as a trout expert, a recognition he deserved. And now James Prosek was being wooed by National Fairways to be the promoter for an exclusive golf club with trout in its name.

I met up with James in mid-July at the Bluebird after he contacted me to say he had just published another book, *Joe & Me*, an autobiographical narrative that chronicled his friendship with fishing mentor Joe Haines, the same BHC game warden I mentioned in chapter 11. Joe Haines, the story goes, caught James poaching in the Easton reservoir when he was fourteen. Instead of arresting him, Haines took James under his wing to show him all the legal streams, lakes, and even reservoirs he could fish. The poacher turned student of the outdoors soon became the teacher, and the rest is angling history.

When I saw James at the Bluebird counter, he was in good spirits, excited about the new book, which this time had human characters, as well as fish. The Bluebird was a good spot to meet since I had scheduled a tour of the Trout Brook property less than a half-mile away at 3 p.m. with *Times* photographer Tom McDonald and Easton officials. When the topic of Trout Brook came up, James's expression changed. He said that National Fairways wanted to commission him to do a number of trout paintings to decorate the walls of the golf course clubhouse. In addition, the developer wanted to name each hole after a

Fig. 35. The planned 11th Tee in Trout Brook Valley. Photo by Paul Avgerinos.

different trout species, and reproductions of James's paintings would be on signs at each hole.

"I told them no, it's a conflict of interest," James said matter-of-factly. "I'm not going to do it, because I don't want to see it developed."

The reason was obvious, considering the very opening to *Trout: An Illustrated History*:

> One of my greatest passions in life is fishing. For me it's more than a pastime or a hobby; instead it's a way of life, and an escape. The environment of the stream is peaceful. The sound of the water over the rocks is soothing, and any sound of civilization is muffled by the trees. This solitude allows me to brush aside daily life and focus of the task at hand of catching a fish. The brook, then, with all its colors and sounds, is an education, a place where I can learn about myself and other creatures as well (Prosek, p. 3).

In short, James's world was the antithesis of what Mr. Bergschneider had envisioned.

Trout Brook was actually a new appellation created by the water company. The watercourse's real name was Hawley's Brook. It begins up north in the marshes of Redding and winds its way down through

Fig. 36. The planned 9th Tee in Trout Brook Valley. Photo by Paul Avgerinos.

Easton to Weston, where it turns behind the ridges of the Saugatuck Reservoir and heads south. Eventually, it empties into the Saugatuck River just below the dam. Near this area off Valley Forge Road there is a side road named Bradley Road. Hawley's Brook, a.k.a. Trout Brook, cuts through the property of a few residents here. Summer evenings, when the mosquitoes and flies hover above its waters, some say the sound of trout jumping becomes so intense it sounds like popcorn popping.

Though Marc Bergschneider had told me that today's eco-friendly golf courses had come a long way from the days of intense pesticide and chemical fertilizer use, it was impossible not to think that links, not to mention the houses, would have an effect on Hawley's Brook. Minor runoff of even the least invasive chemicals would affect the food chain in the brook because the trout would have less larvae to eat. That popcorn sound would cease.

After lunch with James I drove up the road and parked near the entrance to the old apple orchard, where I met photographer Tom McDonald, Gail Bromer of Easton's Conservation Commission, Princie Falkenhagen of Citizens for Easton, and Curtis Johnson, an attorney for the Connecticut Fund for the Environment. Introductions were exchanged, and I learned we would have to wait for a water company van so we could tour the property. The driver, his name escapes me now, seemed friendly enough at first as we headed up the dirt road to a field that had been the orchard. But as my questions about their plans to save the land from development and its effect on Easton continued, the BHC driver appeared to be getting aggravated. The vehicle stopped, and we got out to walk the old orchard. It was unrecognizable. I tried to remember where the apple trees had been, or the fields, but I couldn't place them. Toward the south end of the cleared area Tom and I ascended a dilapidated, gray structure that resembled a poorly designed lookout tower or hunter's platform. From up there, the views were unencumbered. In the distance you could see the blue waters of the Long Island Sound, in addition to the Aspetuck and Hemlocks reservoirs. The Saugatuck Reservoir, however, was blocked by a western ridge. Tom snapped photos; I kept looking around and asking questions, trying to figure out if this, about the highest point on the property, would be the site of the clubhouse, the housing development, or the fairway. No one was quite sure.

In a short while, we were back in the BHC van for a roundabout, dirt road journey into the woods. We passed fields, old stone walls spotted with moss and lichen, woods that seemed to dip down into a natural, dark basement, hillsides of mountain laurel and wild blueberry bushes, and wetlands where the sun peeked through and the dirt road became puddled. We were heading toward the Weston side of Trout Brook Valley, and when Tom and I asked if we could get out to explore, the driver was happy to oblige. He left in a hurry, as if he had to show the others something more important. I began to feel as if we had been ditched. Not that I minded much. Despite the humidity, the bugs were not that bothersome. And it was a quiet woodland walk, the kind of Connecticut outdoor retreat that seems hallowed as you plod the trails and narrow cart paths. I could envision settlers guiding yoked oxen, farmhands straining with ploughs and in stoic acceptance of each stone they unearthed, a fact-of-life harvest they would pile elsewhere and eventually use to build stone walls. The walk through these woods made me wish that just for a few minutes I could trip the time-space continuum and go back to a presumably less complicated and perhaps more innocent time where everything would exude the same warm, homey feeling found in old photographs.

As I've said, the Valley Forge story was pretty much unknown to me at this point. I once heard it said that some houses were flooded to make the reservoir, but it hadn't really registered. Now, I was walking through the remaining woods that had been its residents' fields. The stone walls that had kept cattle in or served as property lines separating Valley Forge's properties now lay like wrecks on the floor of a green sea and entangled by bittersweet, poison oak, and clusters of wild blackberry bushes. I stumbled, literally, across low-lying stone foundations of barns and small homes. To the sides of them lay nearly petrified, rock-hard and rough-hewn beams looking shotgun blasted by years of termites and wood-eating larvae. I wondered what life was like when these beams held up a home. A tremendous respect for the old farmers, whatever time period they came from, consumed me. How could one not respect those whose backbreaking toils carved out a community in this rocky, often inhospitable terrain? It wasn't just survival that had stoked their efforts. They had been driven by a sense of permanence in this place, grounded by a belief that the land they had tamed was their legacy to their children's children and beyond.

In the distance we could hear the BHC vehicle approaching, and Tom and I walked out to the dirt road to flag it down. The driver slowed, and as we approached it he shot off leaving us in his dust. Tom and I mouthed something about the driver and continued walking in the direction of the van. Before long we were at the end of the road and on the paved section of Bradley Road in Weston. We managed to get a ride back to our cars in Easton from Princie Falkenhagen. Hot, sticky, and annoyed, I needed a cool drink.

14

Into the Woods, Again

It was mid-September, the leaves had not yet turned, and each time the sun pierced the boughs, swaths of gold cut the mist from the morning rain. It was a form of natural stage lighting, an interplay of light and dark, a shaded intrigue found only in the deep woods. A melange of sweet, tangy and pungent aromas rose along the trail. It was the second time I would walk Trout Brook Valley. This time it was with my son's Cub Scout troop. We had entered by a dirt road off Route 58 in Easton near the Redding border.

I walked ahead of the troop and saw warblers take flight. A salamander darted off the trail, and a couple of Northern frogs hopped aside to hide under a skunk cabbage. These were wise moves as the scouts, like a wild herd in yellow rain slickers, ran and stomped down on everything in the path, leaves, twigs, and mosses. They kicked small stones, assorted fungi, and whatever else amused them.

"Slow down and listen up," yelled the scout leader. Then he motioned to Mark Harper, a bear of a man who was leading the day's hike from Redding through Trout Brook and around the Saugatuck Reservoir.

Mark Harper, with a long, untamed beard tinseled with gray, wore camouflage fatigues and a hunting cap squeezed down over his long matted hair. He looked as if he had just walked out of Whitlock's photo of Valley Forge's hunters. He was at home here. His eyes, which often displayed the glare of a biker with an attitude, seemed uncharacteristically gentle in the woods. This wasn't Mark Harper the blustering animal control officer for the Town of Weston, or the Mark Harper who boomed about environmental issues and taxes at town meetings. Instead, he displayed the presence of a man overwhelmed by a tremendous respect for where he was. Reverently, he spoke about the woods the same way a new father would describe a sleeping infant, bursting with pride but not wanting to wake the child.

Maybe it was the mist that made Mark's eyes moisten, but there was no denying the sentiment. He described how his father, Robert, an avocational archaeologist, would take him as a boy on digs in these woods, expeditions that netted hundreds of Native American artifacts now on display at Yale University. This was also the outdoor classroom where young Mark had learned to hunt and fish and had been schooled never to take more than was needed.

"Right here we have an Indian cliff dwelling," he said point to a towering ledge. It seemed a palisade of more than two stories. The scouts squinted as they looked at Nature's own high rise, an ancient coop for people who learned to coexist rather than conquer the landscape.

Fueled by juice boxes and snack bars, the troop moved on into an amusement park of trails, foundations, hillocks, jagged ledge, streams, swamps with sedge clusters like stepping stones, and mud, lots of it, the kind that made slurping sounds when the boys stomped into it.

As the day progressed, the hikers wound their way to the Weston border where the forest's dark covering was replaced by brightness in the distance. From afar it looked like a clearing, but as they moved closer the boys could see the path opening up onto the great expanse that was the Saugatuck Reservoir. From this vantage point the reservoir could have been a major New England lake. The troop decided to rest. Some boys sat down on the boulders. Others found places to rest along the dry section of stone wall, the part that didn't look like it had been painted with mud as it rose out of the water. After a fifteen-minute break, Mark directed the scouts forward. They walked for a seemingly endless time around the shoreline before entering the woods again by way of a trail with a hidden opening.

Mark wanted the boys to absorb all they could about the forest, and there was almost a sense of urgency about his educational tour. Though he had led this hike every year, each time with a new batch of scouts, this outing was different. He acted as if he wanted to savor each step, perhaps for the last time. There was a growing, sad realization that much of this unspoiled acreage might not be there in a year or two. And near the end of the hike, the writing was on the wall, or, more specifically, on a tree trunk. A maple had been spray-painted with the number 17 denoting that this was where the seventeenth hole would be. A quiet swept the troop that had just spent the day falling in love with the woods only to learn this playground was earmarked for the bulldozer. One would have to love golf more than the woods to let that happen, they figured, love golf more than Tiger Woods even. The boys began talking and playing again.

"Listen up, Scouts," the troop leader yelled. He raised his hand showing two fingers together, the scout sign for quiet.

"Over there," he said pointing to a downward slope, "is where they plan to build a golf course. "And unless you have $100,000 to join, this might be the only time you get to see this land."

"Can't somebody do anything to stop this?" I heard one father ask.

History and changing demographics seemed to say no.

Little did I know that at this time in Easton a grass-roots movement to save Trout Brook was taking form. Louise Kranyik, one of the Easton residents I interviewed, was also a longtime member of the Easton Homemakers, a local women's organization that offered members programs of interest. She had been charged with developing a program for an upcoming meeting. She was at a loss for a topic when her husband, Bob Kranyik, presented her with the obvious. He pointed to my story, "Pricey Houses, Golf in Easton's Future?" on the front page of the *Times*' Connecticut section Sunday, August 31, 1997. Louise had received many comments at the Bluebird and around town about the article. And people were asking what could be done to save the valley.

Bob Kranyik suggested Louise contact Curtis Johnson from the Connecticut Fund for the Environment to be the group's guest speaker. Though Mr. Johnson replied that he could not attend, he said that the assistant director, Carolyn Hughes, would.

Around the same time, Princie Falkenhagen, president of Citizens for Easton, had made Trout Brook the focus of the group's monthly

Fig. 37. Cover of Trout Brook Valley brochure. Courtesy of Coalition to Save Trout Brook Valley.

board meeting. She began a campaign of writing letters to the editor of local newspapers in which she stressed that Trout Book Valley was not a done development deal. She and Gail Bromer also prepared a flyer on the need to save the land. It was the beginning of a publicity campaign that would ultimately make national news. Bob Kranyik wrote that "volunteers passed out the flyers at the polls on Election Day to alert Easton citizens to the development, its implications to Easton, and the fact that it was not 'a done deal'" (p. 39).

The citizens group also agreed to collaborate with the Easton Homemakers on their November 11 meeting in the social hall of Notre Dame Church.

"'The fate of Trout Brook Valley is a statewide issue, not just a matter of concern for Easton and Weston,' an official of the Connecticut Fund for the Environment declared at a public meeting in Easton Tuesday night," writer Harold Hornstein reported in the November 14 issue of the *Westport News*. "The CFE official, who works on open space and water company land issues, advised approximately 75 people at the meeting to act quickly to head off the proposed development."

Mr. Hornstein reported that the event resulted "in a rousing display of support for preservation." According to Bob Kranyik, the meeting his wife orchestrated was "the first tangible indication that Eastonites (and some Westonites who attended) were concerned about the possibility that some very special open space might be lost forever."

What followed was the creation of the Coalition to Preserve Trout Brook (a.k.a. Coalition to Save Trout Brook a group many at first considered quixotic, a coalition reminiscent of the Saugatuck Valley Association to those who remembered it. Much of the fight to save Trout Brook, in fact, resembles the Valley Forge saga, as if the tunnel connecting the two reservoirs were a wormhole to an alternate place and time.

There was that same underdog determination to save a valley. That same support from cultural and political leaders. That same cause célèbre. Only this time, the water company's bald-faced claims about the project being in the public's best interest didn't wash. How much of a necessity was exclusive, high-end golf? And the argument about the land sales being used to offset rate increases seemed a stretch, too. Eminent domain was not an issue this time, either. The land had been the water company's property for sixty years. Still, the question about

PRESERVATION UPDATE

Volume 1, Number 1   Spring/Summer 1999

Dear Friends and Supporters:

We are just shy of a year since Aspetuck Land Trust's momentous signing of the agreement to purchase Trout Brook Valley for open space. In the meantime a large sum of money has been raised through private donations and fundraisers. Many people have had opportunities to walk the Valley, and countless hours have been spent on preliminary planning for use of the property as open space.

A new organization, the Friends of Trout Brook Valley has been created. It functions as a committee of the Aspetuck Land Trust, assists in on-going fundraising and helps to plan, finance and carry out programs for the Valley. Anyone interested in working with the Friends should contact Princie Falkenhagen at 203-261-5490.

Support for the Valley has been overwhelming, and for those who have been here from the early days, we offer heartfelt thanks. This newsletter has been produced to bring you up to date and to keep you abreast of the progress which has been made as we look forward to the closing on Trout Brook Valley in only three months, on or about September 5, 1999!

Aspetuck Land Trust Report

One June 5, 1998 the Aspetuck Land Trust exercised a right of first refusal as provided by state statute to purchase 668 acres of Trout Brook Valley from the Bridgeport Hydraulic Company (BHC) for $12.4 million. We currently have firm commitments from the State of Connecticut for $6 million and the town of Weston for $845,000. Over the past year, The Nature Conservancy, Aspetuck Land Trust and the Friends of Trout Brook Valley have raised over $3 million privately. We are left with a $1.9 million balance to raise. Though The Nature Conservancy has guaranteed to bridge any amount we are short at the time of closing in order to go forth with the purchase, we will continue to fundraise to repay that amount to the Conservancy as quickly as possible to spare the interest charges on the loan.

Hopefully you have already received a flyer from the Friends informing you that Senators Dodd and Lieberman and Representative Maloney have requested $2 million in federal funds for the preservation of Trout Brook Valley. The Connecticut delegation has a few other large acquisition projects it is working on also, so there is great uncertainty over the total amount of open space funds which will be approved by Congress. Please write your congresspersons and let them know you support their efforts on behalf of Trout Brook Valley.

The Nature Conservancy and Trout Brook Valley

Valley's Significance Placed On A World Wide Level

Nature Conservancy President John C. Sawhill recently featured Trout Brook Valley in a national mailing to supporters, recognizing its significance in a broad conservation context. The Valley was identified along with special parcels of land in Papua New Guinea, Brazil, Jamaica, Hawaii, and Oklahoma. The Connecticut Chapter has committed to helping protect Trout Brook Valley for reasons closely tied to its mission of protecting biological diversity.

- The Valley lies withing the site design of Devil's Den Preserve, meaning that protecting it will enhance the conservation work the Conservancy now pursues at the Den.
- The 668 acres contain nearly all of the Hawley's Brook watershed. The brook is a stream of exceptional quality largely because it has been protected as a watershed for more than a half a century. A stream of such high quality is likely to harbor aquatic invertebrates that are intolerant of disturbance, although an inventory of such species has not yet been conducted.
- The Valley is characterized by a mosaic of oak, hickory, beech and maple forest and supports an array of species that characterizes the representative natural community of the lower New England Ecoregion.
- The forest supports numerous raptors, including the goshawk, sharp-shinned hawk (listed as endangered in Connecticut) and Cooper's hawk (a threatened species in the state) that prey on smaller birds. The abundance of these raptors, all of which require large tracts of forest, suggests that the forest that includes Trout Brook Valley is relatively undisturbed, and supports sustainable populations of the migratory birds upon which these larger birds feed.

Fig. 38. Cover of "Trout Brook Valley Preservation Update," Spring/Summer 1999. Courtesy of Coalition to Save Trout Brook Valley.

the greater good did resurface, and this time the drumbeat came from environmentalists.

The Coalition to Save Trout Brook brought together a creative brain trust for fund-raising and publicity strategies. Along with Princie Falkenhagen, Gail Bromer, and the Kranyiks were Arianne Tallman, who lived on the Weston side of Trout Brook Valley; journalist Verne Gay; writer Dean Noble; and a host of volunteers including James Prosek, Dina Elliott, Janey Hoyt, and Dick and Judy Richardson. Dean Noble is credited with the slogan that would be advertised across the state for nearly two years: "Trout Brook Valley. Forever Yours. Or Forever Gone."

15

Celebrity Status

The Coalition to Save Trout Brook had initially assumed it would need the full $14.2 million to purchase the valley. It had hoped the state would come up with half, and that Easton and Weston would fund the other $7.1 million. To a lot of Easton residents, the coalition's goal must have looked like delusions of grandeur. Raising the necessary funds was a very long shot, especially with Easton divided over whether or not development would be beneficial to the town.

Meanwhile, National Fairways took very logical, well-planned steps in the hope that golfers would be teeing off within a year. The company had already submitted its initial development plan to Easton's planning and zoning Commission in October. Marc Bergschneider, obviously, could not have expected a quick approval. After all, his was not the usual deck, house addition, or new home construction application that the commission saw. In fact, this 200-acre, 7,000-yard championship golf course with 21,000-square-foot clubhouse and 103 super-large homes in a gated community was the largest project the commission had ever faced. Meetings were destined to be involved, heated, and contentious.

This was especially true as the Coalition to Save Trout Brook accelerated its public relations media machine. Verne Gay and Dean Noble were netting results. No longer was talk about Trout Brook confined to gripe sessions at the Bluebird or Weston's equivalent meeting spot, the Lunch Box, or to letters to the editor and stories in the *Easton Courier* or the *Weston Forum*. Interest spread to the *Fairfield-Citizen News*, the *Fairfield Minuteman*, the *Westport News*, the *Westport Minuteman*, the *Connecticut Post* and wire service stories.

These articles caught the attention of other open space advocates, among them Connecticut governor John G. Rowland, who had made the preservation of open space a centerpiece in his bid for re-election. Rowland had already established a Blue Ribbon Task Force on Open Space in August 1997 to conduct hearings around the state to look at as many as 110,000 acres that should be preserved. Governor Rowland fervently believed land preservation was key to protecting Connecticut's natural identity and quality of life.

Still, Easton was divided over Trout Brook, and surprisingly, the topic was barely mentioned in the race for first selectman in November. The reason was that the first selectman candidates William Kupinse and Robert Lessler had agreed to limit discussions after a meeting at the Bluebird with Mr. Bergschneider, Mr. Colonnese, and other members of Mr. Bergschneider's development team. This fact was reported in the November 19 issue of the *Fairfield-Citizen News* and later recounted in Robert Kranyik's book. The newspaper article written by Harold Hornstein and Debra A. Estock won first place for environmental reporting from the New England Press Association.

Robert Kranyik recalled, "the article caused a furor in some quarters. In retrospect, however, the move may have been motivated by the notion that development would be a good thing for Easton since it would promise tax revenue to the town. Nevertheless, it was a precursor of other undercurrents which would affect the effort to preserve Trout Brook Valley, several of which would surface in the months to follow" (Kranyik, p. 44).

Simply put, despite the groundswell of support the coalition was getting, there were still those in Easton who favored development.

But in December, a number of events would work in the coalition's favor. One was a standing-room-only meeting of the governor's Blue Ribbon Task Force held at the Easton Library thanks to the prompting of Gail Bromer of Easton's Open Space Task Force. Then another state body became interested, the Department of Public Utility Control (DPUC), to which the water company answered. The DPUC held hearings in Weston and Easton that month to discern the effect the development would have on drinking water, customer rates, and nearby residents.

The coalition also received its first editorial endorsement from the *Fairfield Minuteman:* "Their cause is a worthy one which residents throughout southwestern Connecticut would do well to heed. That is because the development of the scale currently envisioned for Trout Brook Valley will affect not only Easton, but the social and ecological balance in Fairfield and other nearby towns" (Dec. 4, 1997).

While all positive signs for the Coalition, none was as important as what occurred at the December 22 meeting of Easton's planning and zoning commission held at the town's Samuel Staples School. Eyes turned at the standing-room-only meeting when screen legend–philanthropist Paul Newman walked in accompanied by his two daughters, Melissa and Nell, and their husbands. Mr. Newman, a resident of the neighboring town of Westport, stood on the side and listened intently.

"Mr. Newman did not address the commission but observed from the side of the Samuel Staples School gymnasium as all but two of about seventeen residents expressed strong opposition to the proposed golf course," Mr. Hornstein reported in the December 26 issue of the *Westport News*.

Though Mr. Newman remained silent, his daughter Nell and her husband, Frank Barron, "strongly opposed the use of the land for a golf course and urged that it be kept in its natural state," Harold Hornstein sated.

Bob Kranyik, who was at the meeting, said that during a break in the hearing, Melissa, known as Lissie, sought out the leader of the coalition and was soon introduced to Princie Falkenhagen. Lissie then asked Princie if she would meet her father, who was waiting in the hall. There, Paul Newman is said to have asked Princie Falkenhagen what the coalition would need to save Trout Brook Valley. Her response was succinct: money.

"That's when Paul Newman responded, 'Would $500,000 help?'"

Bob Kranyik recalled. Newman suggested that his philanthropic organization, Newman's Own, might even be able to do better if there was enough public support.

Princie Falkenhagen and coalition members were ecstatic, and word of the Newman pledge circulated among those at the meeting, though no formal announcement was made.

"I just don't think a golf course is the proper use for land that was bought for the single purpose of a reservoir for a growing Fairfield County," Mr. Newman was quoted saying by Harold Hornstein. "Why would they buy it for one purpose and then change it? It's shortsighted." He also bemoaned the fact that Connecticut ranked next to last in the preservation of open space among eastern states despite the fact that it is the richest state per capita. "You'd think at least 500 families in Fairfield County would contribute twenty thousand bucks, which could be deducted from taxes."

According to Bob Kranyik, the Newman pledge "placed the effort to save Trout Brook Valley into high gear, by suggesting that raising the funds would be possible." The pledge would provide $100,000 over five years to the Aspetuck Land Trust, which had the right of first refusal.

"The Land Trust was the bank, and all checks would be made out to them," Princie Falkenhagen points out. "We were just the fund-raising arm, educational arm, and the grass roots in charge of fomenting information about this."

And getting that message out became easier once Mr. Newman came on board. Saving Trout Brook, in the public eye, became a noble, respectable cause. After all, here was a man whose charity funded by the sale of Newman's Own food items had supported many worthy causes. The Hole in the Wall Gang Camp in Stafford Springs, Connecticut, was a prime example. Newman's charity had purchased more than a hundred acres in this northeastern Connecticut town and had transformed it into a summer camp for terminally ill children. Such largesse positioned him as more than just a legendary actor. This was Paul Newman the man with a social conscience, the one whose foundation had already given more than $100 million to different charities. He was, to harken back to the Valley Forge era, a 1990s version of Lillian Wald.

But would celebrity and the national spotlight be enough to win against big business and a public utility? History and dollars appeared to favor BHC. Bob Kranyik, for one, had studied the story of Valley Forge, and he knew that if Trout Brook Valley were to be saved, coalition members could not afford to repeat the mistakes of the past. They could not tire nor ever let themselves believe theirs was a lost cause.

And despite the coalition's added star power and money, National Fairways and BHC remained determined. That's because Easton's planning and zoning board, not Paul Newman and his coterie of environmentalists, would make the ultimate decision. And again, there were those in Easton who favored the development as a boon to the tax base.

But the day after Newman's pledge, Trout Brook Valley was no longer a local issue. On December 23 the state threw Marc Bergschneider a curve ball as the Department of Environmental Protection stepped in to review the full scope of the project. The DEP had serious concerns about the development's effects on inland wetlands and tributaries that would find their way to the drinking water.

"The DEP representatives also mentioned that this project would, if undertaken, be the second largest ever built in state history," wrote Robert Kranyik (p. 51), observing that the DEP's involvement caused some consternation among those who feared Easton was surrendering local control to a state agency. Still, there were those who wanted to save the valley who saw the state getting involved as a good sign. If anything, it would put more obstacles in Mr. Bergschneider's way.

16

Letters to the Editor

Becoming the editor of the hometown newspaper is something many seasoned journalists do at the end of their careers, sort of a segue into retirement for print junkies. So when I was offered the job in mid-December 1997 to become the editor of the *Westport News* following Woody Klein's retirement, I had some reservations. I was too young to "retire." "Just do it for a year," Christine, my wife, said. "Every reporter dreams of being the editor of the hometown newspaper." Okay, a year, I said. That year turned into a two-year wild ride fueled by a passion for issues that was uniquely Westport. Being editor gave me a front-row seat in a contentious town where everyone had an opinion they weren't shy about sharing. Thus, the paper was never at a loss for news or letters to the editor, and when it came to Trout Brook, the letters were many even though the valley was in another town.

Among the first was a letter from Dr. Albert S. Beasley, a Westporter eager to thank Mr. Newman for his pledge. "One and all, we must raise our pens and voices to save what little undeveloped land remains in our communities," he wrote in the January 4 issue. "If we are unable to preserve what we have now, it will be detrimental to our habitat and will lessen our quality of life.... Paul, I thank you, and I hope that more of us will support the need for open space as best we can. They're not making it any more, and when it's gone, it's gone forever."

In the January 9, 1998, issue of the *Westport News,* Princie Falkenhagen again tried to drive home the regional importance of saving Trout Brook Valley.

"Easton's most important natural resource is open space which ensures its purity of water supply, its visual beauty, wide open vistas and farmlands," she wrote. "Just as we are dependent on neighboring communities for our jobs and commercial needs, so the area around Easton is dependent on Easton for its clean water supply, Christmas tree farms, orchards and open spaces. It is, therefore, our regional responsibility to preserve and protect these natural assets."

She concluded with a plea to Easton residents to come up with the necessary funds to help purchase the land. "It is now our turn to become the stewards of this land and preserve it for future generations."

In that same January 9 issue of the *Westport News* Richard K. Schmidt, chairman of the board for BHC, received equal space. He wrote that he was compelled to set the record straight. Mr. Schmidt stressed that claims that the development would adversely affect the region's drinking were false, because "BHC's first and foremost mission is to provide high quality drinking water to about 500,000 people in the 26 towns that comprise our service area.

"We will not do anything to compromise the safety and quality of the water we supply to our customers," he went on, "In fact, we perform 120,000 tests annually to ensure that quality."

He then explained why these Class III lands were no longer necessary for BHC to keep.

"Having land has been part of our business since BHC was first founded in 1857 as the company built its water supply system," Mr. Schmidt wrote. "As large parcels of Class I and Class I watershed land were bought, some of the acreage associated with these purchases was not needed to protect our water.... This surplus land is called Class III, off-watershed land and does not contribute to the quantity or quality of our drinking water. Trout Brook Valley falls into that Class III category."

He didn't explain why the land had been purchased in the first place. Or why owners had the hardship of parting with their property for low prices in the shadow of eminent domain. Or why the water company hadn't only acquired land that was absolutely necessary. In the 1930s Samuel Senior had emphatically affirmed that none of the land was obtained for the purpose of speculation. But that was exactly what BHC was doing, reaping the benefits of prime real estate they had acquired decades earlier.

According to Richard Schmidt, the sale of these lands now was in the public's best interest, since it would pour more money into water company coffers to offset the need for rate increases.

"Advocates [of open space] would have our customers and shareholders pay," his letter continued. "We believe this is unfair. A majority of our customers reside in inner city environs, specifically, Bridgeport and Stamford. We talk to them every day, and I can assure you that increasing their water bills to pay for open space is not high on their priority list.

"Acquisition of open space is a complex local and state issue," he continued. "If it is high on the priority list of the citizens of Connecticut, we should all participate in funding its acquisition. To single out small groups, such as water company constituents, unfairly places the burden on a small group of citizens, many of whom can not afford to pay for open space."

In other words, Easton and its affluent surrounding towns were being selfish by wanting to save open space when poorer cities needed drinking water and water bill rate relief. It was a masterful spin. I couldn't get over how Mr. Schmidt was trying to make struggling locals feel guilty that they would somehow hurt the poor by refusing to let pristine land be destroyed for an obscenely expensive, exclusive community that would make BHC $14.2 million richer.

Meanwhile, even with Richard Schmidt's assurances that the BHC's drinking water would be protected, there were strong concerns about the effects of the planned development on Hawley's Brook and the Saugatuck River below the dam. There would be runoff of silts, fertilizers, herbicides, insecticides, and salt and sand used on the service roads.

"The construction, cutting of trees, clearing roads and lot sites will seriously increase the amount of runoff on an already stressed river basis system," Margaret P. Gragan, president of the Easton Garden

Club, wrote in a letter to the editor published in the *Westport News*, January 16, 1998. "Not only will this increase erosion in and around Trout Brook Valley, but also will add silt and other debris to the water, possibly clogging the gills of fish and suffocating smaller, high-oxygen–demanding aquatic organisms that form the bottom of the food chain.

Ms. Dragan described the valley as a "critical habitat for interior forest birds which have declined 42 percent in Connecticut since 1972 because of the loss of large undeveloped tracts which they need to survive." She called Trout Brook Valley "the largest un-fragmented forest in Southwestern Connecticut," and said that with adjacent Aspetuck Land Trust and Nature Conservancy property comprised ten square miles of wildlife habitat. Her pitch was eerily reminiscent of the garden clubs that rallied in the late 1930s to save Valley Forge.

"The Easton Garden Club feels so strongly that Trout Brook Valley should be preserved that we are pledging a donation of $10,000," she wrote. "We urge other garden clubs, civic and environmental organizations, and individuals to donate to the preservation effort. Remember, a thing of beauty is a joy forever. Let us keep that joy in the beauty of unspoiled nature and pass it on to future generations."

On January 23, I entered into the fray with the first of three editorials supporting the coalition. Titled "Why Westporters Should Care About Trout Brook," the editorial was aimed at getting more people to think beyond their town's borders. It was accompanied by an editorial cartoon, the idea for which I brainstormed with our cartoonist, Dick Hodgins. It depicted a wide-trunk tree tagged with a sign that stated "Planned 9th Hole," and golfers in the distance. The balloon over one of the two scurrying rabbits read, "I like the hunters better."

The editorial stressed that local control does not preclude shared concerns—especially when it comes to a quality of life issue as important as the environment. "Our natural resources plan a key role in making Westport and Weston desirable places to live in the first place," I wrote. "That tends to be forgotten in the shadows of new-found prosperity, mega-homes and an over-emphasis on affluence that disregards environmental assets. It's the old killing the goose that laid the golden egg.

"Westporter Paul Newman understands this and should be lauded for his charity's $500,000 pledge to help conservationists buy Trout Brook," my editorial continued. "He was wise to take his crusade to the Legislative Office Building in Hartford on Wednesday [January 21, 1998]. There he asked for state aid, as well as the largesse of other philanthropists to help buy the land."

Newman had been accompanied to Hartford by his daughter Lissie and Broadway actor-director James Naughton, a Weston resident who lives near the reservoir property. Perhaps it was more than star power that grabbed state officials' attention. The state had already set the goal of acquiring another 100,000 acres to add to its existing store of 210,000 preserved acres. That goal had become law in June when the General Assembly passed Public Act 97–227. It established that the state would set aside not less than 10 percent of its land mass as open space.

Charles S. Harris, president of the Saugatuck Valley Audubon Society, mentioned this fact in his lengthy letter to the editor that also ran on January 23, 1998. The Saugatuck Valley Audubon Society—an interesting appellation since the Saugatuck Valley is under the Saugatuck Reservoir—is a local chapter of the National Audubon Society, with more than one thousand members locally. Their official statement about the planned development of Trout Brook Valley stressed that it would be "the imminent loss of an irreplaceable tract of open space land."

"At a time when Fairfield County residents find themselves overwhelmed by increasingly congested housing, the preservation of open space has reached a critical turning point," he wrote. "It seems ironic that watershed lands set aside for the benefit of the public are now being sold to benefit a very narrow, elite segment of society."

Harris also pointed out that little was known about the wildlife, "and whether, for example, that wildlife includes any endangered or threatened animal or plant species." Environmentalists had been left in the dark, he said, because the BHC denied public access to these lands throughout the years in order to protect what was considered critical watershed.

"Even more importantly, this is but one example of numerous tracts of watershed lands throughout the state of Connecticut which are for sale, or in the process of being sold," he added. "For the above reasons, the Saugatuck Valley Audubon Society believes that our state Legislature should immediately intervene by establishing a moratorium on the sale of all watershed lands in the state of Connecticut until the effect of the potential loss of these lands can be properly assessed."

His comment was almost prophetic, since the state would eventually intervene two years down the road to save eighteen thousand acres of water company land throughout the state. Meanwhile, the interest in saving Trout Brook grew. It was all deja vu of the Saugatuck Valley Association and the fight to save Valley Forge.

"I was impressed with your editorial on Trout Brook and the challenges for us to forget parochial views and become involved in saving open space, even if it isn't actually in Westport," Roy M. Dickinson wrote in a *Westport News* letter to the editor published January 28. "You say it is in our backyard, and maybe now is one of those times to say NIMBY ["not in my back yard"]."

He added that with Paul Newman's pledge and with Weston's First Selectman George Guidera negotiating for his town to purchase the Trout Brook portion in Weston, "this leaves us with about $12,700,000 to go."

"Now if we could only get all 25,000 Westporters to donate $1.39 a day for the next 365 days, we could save the wetlands, the flora, the fauna, and maybe the Long Island Sound to boot," he added. "I hereby pledge my share. How about the rest of you out there?"

That same day the newspaper's receptionist told me a letter had just arrived that I'd surely be interested in seeing. It was from actor Robert Redford, who owned a home in Weston.

"As a part-time resident of Weston, I am grateful to my neighbors for making me aware of the Trout Brook Valley situation," he wrote in a letter published January 30. "My first reaction to learning that it may be turned into a golf course and housing development was, quite frankly, disbelief. Over 700 acres untouched for many decades, resting in the middle of two protected areas measuring 10 square miles—a golf course and a luxury home development? It just doesn't compute. And I guess that's the whole point."

Like Mr. Newman, Mr. Redford had no reason to grandstand. He was motivated by the fact that he knew the land, knew Hawley's Brook, and as someone with a deep appreciation for the outdoors, wanted it to remain unspoiled. Redford had also made a personal donation to the Coalition to Save Trout Brook Valley, but that amount was not publicized. Robert Kranyik quipped that it was Butch Cassidy and the Sundance Kid teaming up again. Redford's involvement, like Newman's, was predicated by logic and passion without pretense. What follows is the rest of Robert Redford's letter in its entirety.

> Environmental, public health and quality of life issues dominate this debate when you really begin to assess the impact of such a proposal. Think what it will be like to have massive amounts of heavy equipment brought in to plow and pave the roads necessary, to bring in more heavy equipment to bulldoze this unique, natural environment. And after all that, there's a golf course and potentially over 100 homes to build. And after the golf course is built around Hawley's Brook, it will be maintained with large amounts of pesticides and fertilizers. Leaching of these chemicals from the golf course, as well as who knows what from the huge septic fields that will come with the houses, could affect Hawley's Brook, which runs into the Saugatuck River, which empties into the Long Island Sound. Nobody can assure us that water quality and thus, public health and safety, will not be affected for all time. And it's hard to imagine the adjoining preserves won't suffer consequences as well. And that's only the beginning.
>
> The dramatic impact on city services and infrastructure, as well as on a way of life known for years in Easton, Weston and surrounding towns will be nothing more than a sad story to share with our children's children in years to come. Who wins here?
>
> I both applaud and support Paul Newman, his family and the countless others who continue to make this issue their priority. This is really one of those times when we all have to think long and hard about short-term gain for a few at the expense of many. Yes, this is an open space habitat, wildlife and environmental issue. Protecting this land is in the public interest, and the benefits of doing so far outweigh the financial gain a few will realize by going forward with this ill-conceived proposal.

Mr. Redford's support could not have come at a better time for the coalition. That same week time Governor John Rowland unveiled a five-year, more than $160 million-proposal to save about 100,000 acres throughout the state for open space. Under Mr. Rowland's proposal, the bulk of the money would go to the Connecticut Department of Environmental Protection to buy land, and about $59 million would be set aside for cities and towns, even water companies, to save open space.

In a *Westport News* editorial, "Trout Brook, It Seems, Has Been a Wake-Up Call," published February 4, 1998, I commented on the Newman donation and remarked,

> [W]hat's most impressive is the speed with which others have stepped up to the plate.
>
> Private citizens have come forward, some pledging their own money. Attorney General Richard Blumenthal recommended the state step in to save the land. Actor Robert Redford, a resident of Weston, saddled up to join his longtime Hollywood crony. . . . Perhaps the biggest show of support came from our own governor.
>
> All involved deserve credit. There is something basic, primal and priceless about unspoiled, open space. This is especially true when it comes to water company land, which so many of us for so long thought was protected, almost sacred, property. We were wrong. Trout Brook, it seems, has been a wake-up call.

The editorial was accompanied by another Dick Hodgins cartoon, this one of two fish in a stream. The balloon over one of the fish read, "Some really big fish are on our side."

Two days later, in the February 6 issue of the *Westport News*, I ran Debra A. Estock's story, "Governor's Task Force Proposes $20 million Annually for Open Space Purchases," which had also run in the *Fairfield-Citizen News*. It described how Governor Rowland proposed to set aside $20 million annually for open space purchases. The money would come from long-term bonding, lottery proceeds, and surpluses of funds set aside for land acquisition. Ms. Estock detailed the final report of the governor's fifteen-member Blue Ribbon Task Force for Open Space, which stressed "the urgency of acquiring open space."

"Connecticut has far less publicly owned open space, either on a per capita basis or as a percentage of our total land mass, than most other northeastern states," Ms. Estock quoted the report stating. "The average Connecticut resident has fewer opportunities to camp, swim, boat, hunt or hike at public facilities, or simply have access to Connecticut's magnificent natural resources. Furthermore, many private lands which have been used by local communities for these purposes are being closed to public use, either by sale to other owners or due to liability concerns."

In the February 6 issue I also ran a joint letter from Bruce LePage of the Aspetuck Land Trust and Ariane Tallman of the coalition. In it, they tried to appeal to Weston residents by reiterating Mr. Redford's environmental concerns. "All of the water from Trout Brook Valley drains out via Hawley's Brook and into the Saugatuck River which flows through Weston," they argued. "There are potential environmental matters that may affect Weston residents if the development is allowed to proceed."

They described the valley containing "towering glacial ridges, soaring hardwood forests, wetland marshes and rushing streams that speak of a New England virtually extinct in Fairfield County." They went on to point out that if the valley were saved, it would give Weston residents access to a huge natural area for horseback riding, trail biking, and dog walking, activities currently not allowed on water company property.

In addition, I ran a letter February 6 from Marc Bergschneider, who felt it was time to clarify the "great deal of hoopla over the Trout Brook Valley project." He affirmed his support for the Aspetuck Land Trust and "open space in general," but saving the valley, he insisted, would do more harm than good to the town and the state.

"It is clear the project will provide revenue to the town," he wrote. "A recent private study by FXM Associates of Massachusetts indicated as much as $1.5 million in positive annual revenue if the whole project were built.

"This is funding the town sorely needs to build a new school, to enhance existing schools, to help pay for the proposed new recreation center, and to upgrade other open space areas like Toth Park," he continued. "The library needs money so that it can provide Internet service and maintain high employee standards. Every dollar that goes into [saving] Trout Brook is a dollar less that can be applied to these essential needs."

Bergschneider also argued that if the state helped to fund the purchase of Trout Brook Valley for a price lower than $14.2 million, it would "reduce the value of all remaining water company land by more than half.

"This means," he continued, "that water companies will have less assets upon which to maintain lower rates and the proceeds of land sales pay for water treatment centers. This will directly and adversely affect the poorer cities, because that's where the water goes, for the benefit of richer towns. Why not just raise water rates to pay for open space? Obviously, this does not make sense."

Perhaps he was rethinking his approach as he could see public opposition welling up like a huge tidal wave, even suggesting that he would be willing to forego building the 103 "McMansions" on the site.

"As an alternative, we are happy to build only the golf course," he wrote. "This will preserve about 500 acres as pure open space. The town will benefit from the tax revenues and the cost offset from building the access road, parking and maintenance of the property after the purchase. In addition, the overall cost of the property will be reduced because the result will be an ideal private-public partnership.

"The leaders of the Coalition do not want to compromise," he continued. "I have repeatedly tried to discuss this option with the Coalition, but it is all or none. This is not a reasonable position given the positive aspects for the proposed development. Open space is not more important than everything else.

"Hopefully, people are listening to all the facts rather than simply getting caught up in the glamour of saving property. Easton has a huge surplus of open space. People come first, especially children who need a good education. People who are densely packed into cities deserve consideration for open space as well. This is an important debate. Please review the real facts, consider the pros and cons thoughtfully, and let your opinion be known."

People were doing just that. Unfortunately for Marc Bergschneider, they weren't the opinions he wanted to hear.

17

Artful Strategies

After the Newman pledge, donations to save the valley began to trickle in from around the country, as well as the state. Princie Falkenhagen said checks from as far away as California, Florida, and Texas found their way to the Aspetuck Land Trust's war chest, "observing that a lot of these were from people who had lived in Easton and Weston and remembered how beautiful it was and wanted to save the valley." By January 15, the coalition had helped raise more than $700,000.

To increase the flow of support, the coalition and the Aspetuck Land Trust together prepared fund-raising packets designed to tap the deep pockets of Fairfield County's major corporations, as well as individuals. "We were out to raise as much as we could, as much as we had to," says Princie. Any one who donated $20,000 or more would receive a signed, matted, and framed trout print by James Prosek, "and what could be a more appropriate token of appreciation for helping to save Trout Brook Valley than a print of the beautiful Salvenius Fontinalis with its mottled back, orange-gold flanks, and blue and red spots" (Kranyik, p. 53).

While well-heeled individuals throughout the region made donations generous enough to net a Prosek print, the region's major corporations were mysteriously silent. "We never got a corporate donation," Ms. Falkenhagen told me. "I don't know. I guess corporations are funny about open space. We even went to GE [General Electric, headquartered in neighboring Fairfield]. They give to a lot of community events, but they did not give any money. Who knows? Maybe they have a lot of golfers on their board."

Fund-raising efforts had to continue, she added, because the price, $14.2 million, was not negotiable. But even though there would be no haggling, there was a way the Aspetuck Land Trust could get the property for less. State Attorney General Richard Blumenthal, who opposed the valley's development, had a plan.

According to Robert Kranyik, "he had asserted that a private golf course does not constitute public open space and that nonprofit land trusts should not pay the same market price for the land that National Fairways would pay the Bridgeport Hydraulic Corporation" (p. 60). Instead, a fair price would be arrived at based upon the capital gains tax breaks that BHC would receive by selling to the land trust. Under these conditions, the amount the coalition would have to solicit for the Aspetuck Land Trust would be considerably less, but still a stretch for a grass-roots coalition and a small land trust that served four towns.

The National Trust for Public Lands had analyzed the situation, and in March of 1998 stated that with tax benefits to BHC the land could be bought for $7.65 million. The trust was willing to give the money as a bridge loan so the Aspetuck Land Trust could buy it. But a month later the BHC rejected that offer as too low.

All along, the Department of Public Utility Control, the state body that oversees the water company, had been studying whether or not the sale of Trout Brook Valley would be permitted under state law. By the end of February, the DPUC gave its approval providing that all profits from the sale would go to shareholders and to reducing customer rates. Robert Kranyik's book recounted that "the judgment also concluded that a golf course is not open space, so Bridgeport Hydraulic Company could not get credit for open space and thus receive preferential tax break for its shareholders" (p. 64). In other words, the water company could count on no tax breaks by selling to Mr. Bergschneider. But BHC could count on substantial tax savings by selling to the Aspetuck Land Trust.

Meanwhile, Princie Falkenhagen and Bob Kranyik remained tireless in their pitch for donations through speaking engagements to various civic groups throughout Fairfield County. By March more than $800,000 had been raised. While major corporations made no effort to support the cause, Ms. Falkenhagen said one local business, the Greenwich Art Gallery in Fairfield, was eager to help by hosting an art auction with all of the proceeds going to the Aspetuck Land Trust's Trout Brook Valley Fund. More than forty local artists donated original paintings and sculptures for the auction held on Saturday evening, April 4. More than $14,000 was raised, and afterward donations continued to come in.

Of course, the coalition and the Aspetuck Land Trust would not have to raise the full amount, estimated to be about $12.2 million with BHC tax incentives, if the land trust were to exercise its right of first refusal. They could already deduct the price that Weston was willing to pay for the forty-five acres in its own town. Weston was now willing to exercise its right of first refusal to buy the land from BHC, instead of acquiring the property from Mr. Bergschneider after the fact. The reason was that National Fairways had rejected Weston First Selectman George Guidera's undisclosed offer as too low. BHC, however, was asking about $20,000 per acre, which would make Weston's tab in excess of $800,000. This price seemed reasonable to Mr. Guidera. He was also looking for ways to help Easton, though he was emphatic that his administration was not interested in buying land in another town.

"We were basically going to buy what we were allowed to buy, the land that was in our town," George Guidera told me during an interview in August 2004. "We were really led to believe nobody would come through with the money for the rest of the land in Easton, and that there would be a golf course there. So we figured, let's buy the piece in Weston, and we'll go from here."

But even that scenario, Mr. Guidera soon came to realize, would not be the best situation for his town. The more he looked at Bergschneider's development plan, the more he understood that the obvious entry to the development would be through Weston along Bradley Road, which is off Valley Forge Road.

"We saw it was the logical entrance to the golf course. We thought they were thinking it was a more prestigious way to come through Weston, and a shorter and easier way than going all the way around to the other side of Easton. Most of the traffic would come from South to North. This was closer to Westport, and this is where all the construction vehicles would come."

Beyond that, the state Department of Health in March 1998 had rejected BHC's application to build an access road from Black Rock Turnpike in Easton. The access road was denied here because it would cross Class I and II lands and put the water quality at risk (Kranyik, p. 68). So, access through Weston was the only logical course, and George Guidera, an attorney, was willing to fight it all the way to the Supreme Court.

Mr. Guidera's resolve sprang from the fact that he was raised in Weston, and despite the changes he had seen was determined to preserve its rural character. During his six terms as first selectman, the town had already purchased more than one hundred acres of open space. "I want to make Weston the greenbelt of Fairfield County," he told me during an interview for the *New York Times* in 1990.

Preserving open space was something George Guidera practiced personally, as well as professionally. An only child, he had inherited the family homestead on Lyons Plains Road, plus dozens of acres in town. "I could sell it all tomorrow, and have so much money I could fly to Paris every day for coffee, if that were something I wanted to do," he once quipped. "But I won't, because I love the land."

His commitment was as rock solid as the stone walls he had personally built to line his home and the house on the corner of Norfield Road and Weston Road where he has his law office. He was a combination savvy lawyer and fighting Yankee willing to get his hands dirty working the land, a formidable combination for Marc Bergschneider. If anyone could make the Trout Brook Valley project languish by being landlocked, it was George Guidera.

But even with such a staunch ally, the coalition's real struggle was to find enough money to buy the remaining acreage in time. On June 5, 1998, the Aspetuck Land Trust's right of first refusal to purchase the property would expire. If this happened, National Fairways would have, if not a veritable walk in the park when it came to purchasing the Easton property, a clear shot at it.

18

Friends in High Places

All along, Fairfield County legislators including State Senator Judith Freedman of Westport, Senator John McKinney of Fairfield, and Representative John Stripp of Weston pushed for state funding at the Capitol to save Trout Brook.

"Watching the growth that was taking place around here, I was getting scared that if this water company land got into the hands of a developer, we would have nothing left that would make this area unique," recalled Judith Freedman. "And having grown up in this area, I wanted somebody else down the road to enjoy all the things I enjoyed. I was thinking of the future, for people yet to be born."

One of her tasks, she recalled, was to convince legislators from urban areas that this land would benefit their constituents, too. "And after they thought about, they realized that, yes, there were people from Bridgeport who go up there to fish, to hike in the woods, to get out of the city."

"And what's important about Trout Brook Valley is that it is pristine," she added. "You can see what it was like a hundred years ago, even a thousand years ago. That's our legacy for future generations."

Along with legislators like Judy Freedman, the coalition also had a friend in David K. Leff, deputy commissioner of the state Department of Environmental Protection. Princie Falkenhagen described him as a tireless supporter of their cause. Another person was Denise Schlener, who headed the Nature Conservancy's Connecticut chapter. The drumbeat continued in Hartford, and Governor Rowland was listening. "He was very supportive of open space from the beginning," Senator Judith Freedman recalls.

Purchasing and preserving the entire valley, however, would become complicated, if not impossible because of ownership stipulations and red tape.

"It's never easy to really explain the Trout Brook Valley land purchase," explained Bruce LePage, executive director of the Aspetuck Land Trust. "There were 730 acres total, of that Weston would buy forty-five acres, and there were ninety acres that we did not have the right of first refusal to buy."

The ninety acres in question were not owned by BHC, but by its holding company, Aquarion, which was technically not a public utility. Therefore, the Aspetuck Land Trust could not exercise its right to buy it under state law and so, according to Bruce LePage, that left 595 acres that we could buy, and the amount we had to come up with was $11.5 million."

The coalition and the land trust expected the town of Easton to contribute at least $2 million, and the state, as it strove to increase its open space holdings, to fund more than half. The rest would come from coalition and land trust fund-raising efforts, plus a loan from the Nature Conservancy's Connecticut chapter.

As the clock ticked toward the June 5 deadline, state lobbying and local fund raising continued. The slogan "Trout Brook Valley, Forever Yours, Or Forever Gone," appeared in ads, direct mail campaigns, and posters throughout the region, as well as the state. And checks continued to pour into the Aspetuck Land Trust.

"We had a number of people who pledged $50,000 over five years," says Bruce LePage, "but the vast majority were small checks. All together we had 1,972 checks that we processed. We even had a little girl who lived near Trout Brook Valley send us an envelope with four one-dollar bills and some change that she said was her Christmas money."

By early June, the Aspetuck Land Trust's coffers held more than $1.5

million. And then Governor Rowland did something Marc Bergschneider and the water company could never have expected the state to do. The governor committed $6 million to the Department of Environmental Protection toward the purchase.

"I think Bergschneider thought, in fact, I know he thought the State of Connecticut would not come in with any money, " George Guidera recalled. "But this was the state putting its money where its mouth [was] for the first time.

"The state had talked a good game in the past, but this is the first time they made a commitment here in Fairfield County,"he added. "They were used to buying land in other parts of the state where it was cheaper. But they did the right thing because this was land that wouldn't be available here in the future."

Bruce LePage thinks Governor Rowland was obviously moved by Nature Conservancy science that showed it was much better for the environment overall to save large parcels than a series of smaller ones.

"The science the Nature Conservancy came up with said it's worth more to the environment to save one 1,000–acre parcel of land than to save ten 100-acre ones," thus setting the stage for the future purchase of large tracts, LePage believes.

Princie Falkenhagen thinks that Governor Rowland was also motivated by political opportunism. "He knew this was a big area of support, and we did have a budget surplus," she says. "But he still deserves credit, because he did have a task force in place, he did have a plan, and he decided to come out and support us. He took a stand. We also knew that we would be cutting a deal with them where we would have to allow hunting on a portion of this property, but all we could do was bargain with them for controlled hunting."

With the state behind them, the coalition and the Aspetuck Land Trust had a shortfall of only $4 million, and, says Bruce LePage, "before we were able to sign the purchase agreement, we had to have an organization behind us that would back us up for the $4 million." That organization was the Nature Conservancy. Throughout the effort the conservancy's Connecticut chapter kept in regular touch with Dr. Steve Patton, who ran the Conservancy's more than 1,700–acre Lucius Pond Ordway–Devil's Den Preserve in Weston. The Nature Conservancy knew that preserving Trout Brook would be vital to its open space hold-

Fig. 39. A hot-air balloon announces the "Trout Brook Valley Lift-Off" at a celebration atop Flirt Hill. Photo courtesy of Coalition to Save Trout Brook.

ings in Fairfield County. Though the conservancy would stand behind the $4 million shortfall, it was assumed that the town of Easton would pay at least $2 million of that amount.

"So after June 6, we had fifteen months to raise the money," Bruce LePage recalls.

On June 16, Easton held a town-wide referendum to vote on the $2 million expenditure toward the purchase. "Shortly after the polls closed at 8:00 P.M., the results of the referendum were announced in the Samuel Staples auditorium, where the voting had taken place. By a margin of 130 votes, the town of Easton had said 'no' to expending public funds to help save the valley. In the largest turnout since the last presidential election, the voters had cast 1,350 "no" votes and 1,220 'yes' votes. Thus, the town of Easton would not be partner in saving the largest available parcel of open space within its borders" (Kranyik, p. 77).

Bob Kranyik called it a bittersweet moment for the coalition members who had worked so hard for so long. They were happy that the land had been saved but disappointed at their own town, which "had gotten a free lunch," as First Selectman William Kupinse had put it. Princie Falkenhagen attributes the referendum defeat to "a tightfisted old Yankee mentality that believed the money would be better spent elsewhere, such as on the schools."

Still, it was a time for celebration, and on September 26, the coalition held a party with BHC permission atop Flirt Hill, the highest spot in Trout Brook Valley and the site of the old apple orchard. "About 200 people showed up on this warm and partially sunny day to bask in that sun and in the realization that Trout Brook Valley would be indeed, 'Forever Yours'" (Kranyik, p. 80).

The party, attended by state officials including DEP Deputy Comissioner Leff and Attorney General Richard Blumenthal, was not difficult to spot. There were hundreds of well-wishers and supporters who milled around the tent for refreshments while a band played. It was punctuated by a colorful, tethered hot air balloon with the words, "Trout Brook Valley Lift-Off."

And while it was evident the valley would be saved through the guarantee of the Nature Conservancy, the work of coalition members was far from over. Robert Kranyik said they were determined to continue fund raising to help make up the $2 million shortfall due to an Easton majority saying "no" in the town-wide referendum.

At this time the coalition had also changed its name to the Friends of Trout Brook Valley. Though its name conveyed less urgency, the group worked hard at soliciting donations for another full year. It was during this period that Paul Newman proved that he was indeed, a friend of Trout Brook Valley. On February 14, 1999, he and his wife, Joanne Woodward, appeared in the A. R. Gurney play, *Love Letters*, at the Westport Country Playhouse, and all of the proceeds from the sold-out performance went toward the land purchase. The romantic, bittersweet play was the talk of the town. Many who saw it mused how Paul and Joanne must have drawn upon their own relationship as their performance readings choreographed a lifetime of letters between artist Melissa Gardner and attorney Andrew Makepeace Ladd III.

Few among the audience know that more than sixty years earlier noted thespians in Westport also used their talents in a doomed attempt to try to save another valley. On September 3, 1937, Arthur Anderson

 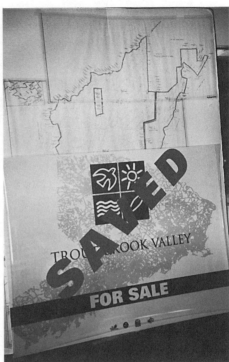

Fig. 40. (*left*) Paul Newman speaks at Weston Town Hall after the purchase of Trout Brook Valley. (*right*) A close-up of the "Saved" poster. Photo courtesy of Coalition to Save Trout Brook.

of the Metropolitan Opera starred in Gilbert and Sulivan's *Pinafore*, held on the lawn in front of the Longshore Club House. All of the proceeds then went to pay for legal fees to stop the Saugatuck Valley and Valley Forge from being flooded.

But Trout Brook Valley was now a "done deal" in favor of preservationists, and any fund-raisers' feelings were far from bittersweet, but colored with optimism and victory.

The actual closing for Trout Brook Valley took place on September 2, 1999, at Weston Town Hall, and the location could not have been more fitting.

"The reason they picked Weston is because we were the first entity that committed to buying the first piece, the forty-five acres," says George Guidera.

So, Weston purchased the forty-five acres within its borders. The

Friends in High Places 131

Fig. 41. Paul Newman and Princie Falkenhagen congratulate each other at the Weston Town Hall closing ceremony for Trout Brook Valley. Photo © 2003 *Connecticut Post* (used with permission).

Aspetuck Land Trust took title to 385 acres in the valley, including the ninety acres that had initially been off limits. That was because Aquarion made everyone happy by first deeding the ninety acres to BHC. The Aspetuck Land Trust also took title to twenty-eight acres outside the valley and, an initial bone of contention, the Four Corners. And the state Department of Environmental Protection took title to the remaining three hundred acres as open space.

"It was a lot of fun, and we had a nice party afterwards," George Guidera recalls, noting how scores of state officials, environmental advocates, coalition members, and celebrities including Paul Newman filled the small town hall.

Amid all the hubbub and cameras flashing, few took notice of a photo exhibition that had opened months earlier in the other hall. Put together by Jim Daniel and the Weston Historical Society, and it told the story of another valley with no such happy ending—Valley Forge.

Princie Falkenhagen was one of the few who took notice of the Valley Forge photo exhibit. She managed to get a copy of an old broadside with the heading, "Save the Saugatuck," and today she has it framed in her home office, where it serves as a reminder that her fight for Trout Brook was not a new one, but one with a historical precedent. It was the same determination the early settlers had when they first carved out communities here, a determination galvanized by a love of place.

"If one valley had to be lost, and another saved, I guess it worked out well the way it did," George Guidera believes. "Otherwise, the water company wouldn't have owned the land, which we were able to get, and all of it would have been developed."

 EPILOGUE

The ink had barely dried on the Trout Brook deeds when local environmentalists sounded a new alarm.

"The British are coming—again," said Woody Bliss, Weston's second selectman who would later become first selectman. It was January of 2000, and his comment had added currency considering we were gathered for dinner near the fireplace in a pre–Revolutionary War home with the Weston League of Women Voters, of which my wife was a member. A self-proclaimed "tree-hugger," Woody was quite animated.

At issue was the fact that the Aquarion Company, the parent of BHC, was recently sold to the Kelda Group in England for $446 million. It seemed like an enormous sum for the purchase of a company that was not that big a revenue producer and in debt. But Aquarion was real estate rich with thousands of Class I, Class II, and Class III lands in Fairfield and Litchfield counties. It was obvious, Woody stressed, that Kelda would want to carve up many of the watershed lands for development, and there were a total of eighteen thousand acres in both counties that were at risk.

"If this property is sold, it will significantly alter the future of Weston," said Bliss, noting that the water company still owned 1,132 acres in Weston. And Easton would fare even worse, with more than seven thousand acres still owned by Aquarion.

In other words, Trout Brook Valley was just a small skirmish in an open space battle that loomed much larger. And if the Trout Brook purchase had seemed an uncertain victory, preserving approximately 18,500 acres would be nearly impossible. At the time I had recently left the *Westport News* to become the editor of *Westport Magazine,* one of those new, glossy, high-end publications geared toward the rich. Needless to say, I missed the newspaper. I was more interested in the struggle of underdogs, not the society pages. But I was eager to bring more issues to the magazine, and the Kelda concerns seemed a good topic for my editor's column.

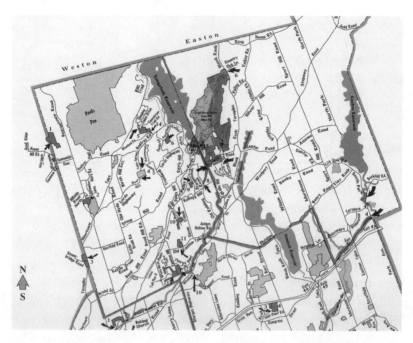

Fig. 43. Map inset: Saugatuck Reservoir and the Trout Brook Valley. Map courtesy of the Aspetuck Land Trust.

"The good news is that state lawmakers, worried about a land grab of historic propositions, have struck a deal with the water company," my column read in the February 2000 issue of the magazine. Under the agreement there would be at least a two-year moratorium on the sale of lands.

"That's so that area towns, cities and state environmental groups can come up with the money to purchase the lands to preserve them," I continued. "The race is on, and hopefully we'll have the same kind of spirit that helped save Trout Brook Valley. The stakes are much higher this time."

The Trout Brook victory had not only galvanized groups like the Connecticut Fund for the Environment and its newly formed arm, the Coalition for the Permanent Protection of Kelda Lands. It proved that the State of Connecticut was serious about preserving open space.

"Kelda knew that the state was now a player in the open space arena, and that the towns were serious about it, too," said Weston's George Guidera.

The possible sale of Kelda lands for development caused a stir in Hartford, and many different avenues for preserving the acreage were discussed. They ranged from buying the lands outright, to buying the lands and/or conservation easements, to the possibility of even creating a regional water authority that would buy the BHC and its lands. The new coalition—co-chaired by Woody Bliss, Don Strait of the Connecticut Fund for the Environment, and Westporter resident Julie Belaga, who had headed the Environmental Protection Agency's New England office in Boston—even proposed acquiring the BHC lands by using eminent domain. If the latter happened, BHC would be dealt the same hand it had used years earlier to acquire these properties.

That June, as I delved into the Valley Forge story to write a feature article on the submerged village for the October 2000 issue of *Westport Magazine*, I saw this last possibility as poetic justice: eminent domain coming full circle to be used against the water company.

The same month I received a press release that Paul Newman was about to hold a press conference in Easton's Toth Park, which is right across from the famed Bluebird restaurant. When I arrived I saw Mr. Newman surrounded by local politicians, area legislators, and Attorney General Richard Blumenthal. Mr. Newman stepped to the podium, and like an angry prophet, he swept his arm taking in the emerald green countryside of Toth Park and its surroundings in Easton.

"The thing I don't want to regret is for my daughter and my grandson not to have this," he said. "It would be regretful if this were a mall, and if citizens would have to back up a quarter of a mile to get to a stop sign. We never really appreciate what we have until it's lost."

He went on to describe Kelda as a multinational corporation based in London, "and there's not a lot of patriotism for this area in London." And he added that someday, who knows what other foreign company may hold title to these lands and decide their future? Perhaps the holding company would be based in Istanbul.

With that said, Newman announced that he had recently joined the Coalition for the Permanent Protection of Kelda Lands and urged others to give their support. He was quickly followed by Mr. Blumenthal who stressed the urgency of getting involved.

"Five years will be too long to wait, and we have an historic opportunity now to make sure our children and grandchildren not see our

18,000 acres lost to development," Blumenthal said. "That will hastily diminish our quality of life."

Their battle plans would be more complex this time, and the Connecticut Fund for the Environment and Fairfield County officials would have to enlist the support of town leaders and legislators in Litchfield County towns, too.

But the groundwork had already been laid with the Trout Brook, which "had become the poster child for open space," Bruce LePage told me. "So they [the Connecticut Fund for the Environment] just increased the lobbying pressure, they created just enough momentum to put pressure on the politicians, and saving open space became the right thing to do."

Governor Rowland had already proven he felt the same way, and the mood in Hartford was that the land should be preserved. Within a year's time, it would be. In February 2001 the governor announced that an agreement had been struck among the state, Kelda, and the Nature Conservancy. Kelda had agreed to sell a combination of lands and conservation easements to the state for $90 million. Bruce LePage suspects Kelda was eager to close the deal to use the money to develop the company's water distribution systems to increase its customer base.

By June 30, 2001, the Connecticut General Assembly approved the final funding package. About $30 million would come from the state's budget surplus, $50 million from state bonding and $10 million from the Nature Conservancy. According to the Connecticut Fund for the Environment, it was the largest acquisition of open space in the history of Connecticut.

"This was comparatively easier because our experience with Trout Brook made us feel that this one could be won," recalls State Senator Judy Freedman. "We learned from the first time that if you work hard enough and get enough people to realize it's important, it can be saved.

"And it fell into place, thank God," she added. "We can't guarantee that we can give people much in the future, but we can certainly give them something like this, something that gives them delight and pleasure by doing nothing to it, by not changing it."

Nothing seems to have changed as I return alone to fish at the Saugatuck Reservoir in early August. There's still that endless green forest embracing the expanse of water under a crisp, azure sky. That first tug of a fish at the end of the line. The quiet as a I walk the rocky shore. Maybe it's the quiet or being alone with nature that makes being here almost spiritual.

I glance back to an empty spot under a tall hemlock where I still hope to catch a glimpse of my father, who since passed away, waving back to me. And other times while I wait, which is what fishermen often do, I try to envision the lives that once thrived below. I want to meet them, like in *Brigadoon,* just for a day. I want to sit in on Florence Banks's class in that drafty, one-room schoolhouse. To net some eels with Clinny Hull. To sit down for Sunday dinner with the Coleys. To hear the sound of a newborn as Doc Gorham with twenty-five dollars in his pocket rides away in his horsedrawn carriage.

And when I think about these things, I'm still bothered by the fact that Valley Forge was lost. Perhaps George Guidera was right. If one valley had to be lost and another saved it was best the way things turned out.

Mary Ann Barr, a trustee of the Weston Historical Society and a co-founder of Keep Weston Rural, put it this way:

"I was just driving around the reservoir today, and I was thinking that yes, it was quite a price that was paid in the loss of people's homes. But being here seventy years later and looking what we have as a result of it, I see it as ironic. If it hadn't been for what at the time was an awful imposition on people's lives, we wouldn't have what we have now. It would all be developed.

"And that thinking might have continued," she adds. "If it weren't for the fact that people enjoyed having all this Bridgeport Hydraulic land over the years, they probably wouldn't have saved Trout Brook and the Kelda lands. I see this string of reservoirs that we have in Fairfield County as an emerald strand, and that is priceless as long as we can keep it. We need open space. We absolutely need it."

This village of the dammed did leave a legacy after all.

Epilogue 139

WORKS CITED AND CONSULTED

"A Great Day for Environmentalists." Editorial. *Westport News.* 10 June 1998: A12.

"Artist Curry Flies East to New York Exhibit." *Westporter-Herald.* 17 December 1937.

"Battle of Valley Forge: Easton, Weston All Join in to Fight Old Hydraulic Foe." *Herald Magazine.* 11 July 1937: 3.

Beasley, Albert. Letter. *Westport News.* 4 January 1998.

Bergschneider, Marc. Letter. *Westport News.* 6 February 1998.

Burlingame, Roger. "Connecticut Spare My Land." *Forum,* 1938.

"Burlingame Scores Water Comany in Bethel Lions Club Speech." *Westporter-Herald.* 14 February 1938: 15.

"Coalition to Save Trout Brook Valley." Pamphlet. 1998.

Daniel, James. *How Things Were in Valley Forge.* Weston, Conn.: Weston Historical Society, 1999.

"Dr. L. I. Dublin Heard in Condemnation Suit of Hydraulic Co." *Westporter-Herald.* 17 June 1938: 1.

Dragon, Margaret. Letter. *Westport News.* 16 January 1998.

Engineer Reveals Statistics on Needs of Hydraulic Co." *Westporter-Herald.* 15 October 1937.

Estock, Debra. Governor's task Force Proposes $20 million Annually for Open Space Purchases." *Westport News* 6 February 1998.

Falkenhagen, Princie. Letter. *Westport News.* 9 January 1998: A17.

Falkenhagen, Princie. Letter. Westport News. 21 Jan. 1998.

Farnham, Thomas J. *Weston: The Forging of a Connecticut Town.* Weston, Conn.: Weston Historical Society, 1979.

"500 Attend Opening of Saugatuck Valley Art Exhibition in New York." *Westporter-Herald.* 17 December 1937: 20.

"Flow of Saugatuck Will Be in the Call Says the Callers." *Westporter-Herald.* 10 June 1938: 1.

"$40,000 Reeived from Hydraulic Co." *Westporter-Herald.* 8 Aug. 1939: 4.

"From the Editor." Editorial. *Westport Magazine.* July/August 2000.

"Gov. Cross to Inspect Devastated Weston Area." *Westporter-Herald*. 23 May 1938: 1.

"Gov. Cross Gives Saug. Valley Assn. His Blessing and Encouragement on Visit Sunday Afternoon." *Westporter-Herald*. 17 May 1938: 1.

Harris, Charles. Letter. *Westport News*. 23 January 1998.

"High Prices Paid for Parkway Land Cause a Stir." *Westporter-Herald*. 24 December 1937: 1.

Hoe, James. "Valley Forge Iron Works." Weston, Conn.: Weston Historical Society, 1999.

Hoe, James. Archival notes. 1999.

Hornstein, Harold. "Newman Offers $500,000 to Save Easton's Trout Brook Valley." *Westport News*. 26 December 1997: A16.

Hornstein, Harold. "Quick Action Urged to Stop Trout Brook Valley Development." *Westport News*. 14 November 1997: A20.

"Hydraulic Company Is Sued for $250,000 by Land Owners Below the Dam." *Westporter-Herald*. 4 February 1938.

"Hydraulic Plan to Be Discussed at Weston Meet." *Westporter-Herald*. 19 November 1938: 1.

"Hydraulic Co. to Halt Cutting of Larger Trees." *Westporter-Herald*. 19 April 1938: 1.

"John A. Anderson Claims Bridgeport Hydraulic Co. Reservoir Will Be an Asset to Weston." *Westporter-Herald*. 7 June 1938: 1.

"Joseph F. Buckley Appeals to Supreme Court of Errors." *Westporter-Herald*. 17 December 1937: 1.

"Judge Wynne's Decision in Favor of the Hydraulic Company." *Westporter-Herald*. 8 October 1938: 15.

Kranyik, Robert, and Verne Gay. *Trout Brook Valley: Forever Yours*. Easton, Conn.: (Self-Published), 2000.

Krieger, Phyllis. "Recalling the 1938 Battles When the Reservoir Expanded." *Weston Forum*. Undated Weston Historical Society clipping.

"Land Revered." Editorial. *Westport Magazine*. February 2000.

LePage, Bruce. Letter. *Westport News*. 6 February 1998.

"Lillian Wald Discourses on the Saugatuck Valley." *Westporter-Herald*. 19 November 1937: 9.

Lomuscio, James. "Pricey Houses, Golf in Easton's Future." *New York Times*. 31 August 1997: CT1.

Lomuscio, James. "What Lies Beneath Valley Forge: Weston's Village of the Dammed." *Westport Magazine*. October 2000.

"Many from This Section Support Bill to Curb Hydraulic's Powers." *Westporter-Herald*. 28 March 1939: 1.

"Many Supported Bill Designed to Curb the Charter Rights of All Public, Private Water Companies." *Westporter-Herald.* 25 April 1939: 1.

"Mrs. Harvey Killed in Auto; Fell Asleep and Car Hit Tree." *Westporter-Herald.* 19 October 1937: 1.

Organization Meeting of the Saugatuck Valley Association Friday Evening at Horace Hurlbutt School, Weston." *Westporter-Herald.* 8 October 1937.

Partridge, Helen, and Francis Mellen. *Easton: It's History.* Collinsville, Conn.: Lithographics, 1972.

"Pictorial Exhibition of Saugatuck Valley Postponed a Week." *Westporter-Herald.* 5 November 1937: 10.

"Pres. Senior in an Address Here." *Westporter-Herald.* 29 October 1937: 15.

"Prices Paid for Land by the Hydraulic Co. Are Aired in Court." *Westporter-Herald.* 2 November 1937: 1.

Prosek, James. *Trout: An Illustrated History.* New York: Knopf, 1996.

Redford, Robert. Letter. *Westport News.* 30 January 1998: 16.

"Redding Rejects Hydraulic Big Offer of $65,000 in Town Meeting Last Eve." *Westporter-Herald.* 7 June 1938: 1.

"Relocation of Some Weston Roads the Next Problem." *Westporter-Herald.* 4 January 1938.

"Saugatuck V.A. Takes Case to Higher Court." *Westporter-Herald.* 5 Nov. 1937: 1.

"Scenic Beauty of the Saugatuck Valley to Be Dramatized Here." *Westporter-Herald.* 2 November 1937.

Schmidt, Richard. Letter. *Westport News.* 9 January 1998.

"Socialists Support Aims of Saugatuck Valley Associaton." *Westporter-Herald.* 7 December 1937.

"S.V.A. Siles Sup. Court Appeal." *Westporter-Herald.* 4 March 1938: 1.

"Saug. V.A. Take Case to Higher Court." *Westporter-Herald.* 5 November 1937: 1.

"Special Committee Agrees on $37,500 Cash from Utility." *Westporter-Herald.* 20 May 1938: 1.

"S.V.A. Sassiety Column." *Westporter-Herald.* 12 April 1939: 19.

"To Seek 1 Million Gallons Water Daily." *Westporter-Herald.* 16 June 1939: 1.

"Town Meeting Was One of Greatest Interest." *Westporter-Herald.* 14 April 1939.

"Town To Consider Riparian Rights." *Westporter-Herald.* 3 June 1938: 1.

"Trout Brook, It Seems, Has Been a Wake-Up Call." Editorial. *Westport News.* 4 February 1998: A12.

"Valley Folks Plead Before Legislative Comm." *Westporter-Herald.* 19 May 1939: 8.

"Valley Forge School District is Formally Abolished." *Westporter-Herald.* 2 May 1939: 4.

"Walter Peck, Real Estate Man Gave Opinion on Weston Land Values." *Westporter-Herald.* 31 October 1937.

"Westonites Taxes Much More Per Acre Than Hydraulic Company." *Westporter-Herald.* 8 October 1937: 12.

"Weston Hydraulic Co. Proposed Plan of Settlement Outlined." *Westporter-Herald.* 20 May 1938: 15.

"Weston Turns Down Hydraulic Co. Proposition for New Roads; $37,500." *Westporter-Herald.* 27 May 1938: 1.

"Westport's Interest in the Flow of the Saugatuck River." *Westporter-Herald.* 14 June 1938: 1.

"W. Waldron Writes About Constructive Measure of the SVA." *Westporter-Herald.* 13 June 1939: 2.

"Why Westporters Should Care About Trout Brook." Editorial. *Westport News.* 23 January 1998.

"Woman Bitten by Copperhead." *Westporter-Herald.* 18 August 1939: 1.

Young, John Orr. Letter. *Westporter-Herald.* 24 May 1938: 1.

INTERVIEWS

Albin, Ernie. 27 June 2004.
Barr, Mary Ann. June, July 2004.
Daniel, James. June 2000.
Falkenhagen, Princie. 5 August 2004.
Franklin, Olive. 10 June 2004.
Franklin, William. 10 June 2004.
Frederik, Burry. July 2000.
Freedman, Judith. 4 August 2004.
Guidera, George. 1 August 2004.
Haines, Joseph. 20 July 2004.
Hoe, Arthur "Jim." 15 June 2000.
Kranyik, Robert. 2 June 2004.
LePage, Bruce. 29 July 2004.
McCullough, Peg. 12 June 2004.
Pida, Julia. 15 July 1997.
Prosek, James. 15 July 1997.
Samuelson, Gary. 10 February 2004.

 **INDEX**

Illustrations are indicated by page numbers in boldface.

Adams, Franklin P., 42
Adelle, Sarah, 22
Adirondacks, 1, 89
Albin, Ernie, 75, 78
Anderson, Arthur, 130
Anderson, John, 56
Anderson, Karl, 52
Angler's Dock, 83
Aquarion, 5, 7, 8, 85, 87–89, 127, 135
Aspetuck Land Trust, 86, 92, 110, 115, 120–125, 127–128, 132
Aspetuck Land Trust's Trout Brook Valley Fund, 124
Aspetuck Reservoir, 9, 78–79, 94, 98
Aspetuck River, 44
Aspetuck Valley Orchards, 91
Atlantis, 3
Audubon, 94
Averginos, Paul, 95–97

Baker, Mrs. John A., 52
Balcolm, Lowell, L., 52
Banks, Florence, 23–25, 66, 139
Banks, Frank, 24
Barnum, P. T., xi, 32, 34,
Barr, Mary Ann, **vi**, 76, 139
Barron, Frank, 109
Beasley, Dr. Albert S., 112
Beers, Hanford, 17
Beers, Steve Harcor, 25
Belaga, Julie, 137
Bennett, Tom, 54
Bergschneider, Marc, 11, 88, 90, 93, 96, 107, 111, 120–123, 125, 128
Berkshire foothills, 14
Black Rock Turnpike, 125
Bliss, Woody, 135, 137

bloomery, 14
Bluebird, 90, 95, 102, 108, 137
Bluebird Garage, 90
Blueribbon Task Force on Open Space, 108, 119
Blumenthal, Attorney General Richard, 119, 123, 130, 137–138
bog ore, 15
Boughton, Mrs. Everett, 64
Boyd, Edward, 52
Bradley Axe factory, 17, 18
Bradley, Charles, 17
Bradley, Emma, 42
Bradley, Hull, 18
Bradley, Sen. J. Kenneth, 50
Bradley Road, 99, 124
Bradley, Wakeman, 17
Brady, Matthew, 10
Brewster, Mrs. John Hull, 42
Bridgeport, xi, xii, 32, 68, 114
Bridgeport Brass, 32
Bridgeport Hydraulic Company (BHC), xi; 3, 5, 6, 13, 31, 34, 35, 37, 39, 40, 46, 50, 56, 57, 60, 62, 63, 65–67, 69, 72, 75, 77, 81, 85, 91, 92, 94, 95, 98, 99, 111, 113, 114, 116, 123, 124, 135, 137, 139
Bridgeport Superior Court, 56
Bridgeport Water Company, 33
Brigadoon (play), 10, 139
Bromer, Gail, 87, 91, 98, 104, 106, 108
Buckley, Aaron, 17
Buckley, Franklin, 28
Buckley, George, 17
Buckley, Joseph, 28
Buckley, Benjamin Franklin, 17, 18
Buckley homestead, **19**

Buckley Machine Shop & Iron Works, 27, 30
Buckley's Mill, **28**, **33**, **35**
Buckley, Nehemiah, 17
Burlingame, Roger, 35
Burr Cemetery, 13, 79, 82

Canaan, 82
Cardini, Linda, 91, 92
Cartbridge Road, 29
Cavanaugh, Mrs. John, 46
Citizens for Easton, 87, 102
Civil War, 3, 10, 16, 33
Clark, Mrs. John Maurice, 64
Class I Watershed, 35, 92, 113, 125
Class III Watershed, 36, 92, 113, 125
Class III Watershed, 92, 113
Clean Water Act, 85
Coalition for the Permanent Protection of Kelda Lands, 136, 137
Coalition to Preserve (Save) Trout Brook, 7, 93, 103, 104, 106, 107, 118, 121, 129, 131, 136
Cobb, Mrs. Frank, 42
Coley, Anna, 76
Coley, Chester, 55, 56, 59, 66
Coley, David, S. 42
Coley, Ted, 76, 77, 80, 83
Colonnese, Anthony, 90, 93
Columbia University's School of Nursing, 47
Connecticut Fund for the Environment, 6, 91, 98, 102, 104, 136, 137, 138
Connecticut Garden Club, 40,
Connecticut Post (newspaper), 108, **132**
Connecticut Rural Development Council, 91
Connecticut's Socialist Party, 54
Copperhead, 67
Cowan, Wood, 40, 42
Cross, Governor Wilbur, 36, 62, 63, 64
Crowe, Bob, 72
Cub Scout, 100
Curry, John Stuart, 54

Daniel, Jim, 3, 8, 10, 12, 14–16, 20, 21, 27–30, 132
Darien, 8

Daugherty, James, 42, 52
Davis, Barton, 51, 54
Davis, Joseph Jr., 17
Davis, Mr., 71, 76
Davis Hill Road, 56, 57
Delafield, E. H., 44
Den Road, 56
Department of Environmental Protection, 111, 119, 127, 132
Department of Health, 92, 125
Department of Public Utility Control (DPUC), 109, 123
Devil's Den, 51
Devil's Glenn, 8, 51
Devil's Mouth, 17
Dickinson, Roy M., 117
Dorr, Mrs. John V. N., 64
Dorr Mill, 64
Dorr, Neil, 52
Dorr, John, 56
Driggs, Fred, 44
Dunning, Phillip, 42

Easton, 1, 9, 35, 67, 68, 79, 83, 86, 87, 89, 90, 92, 97, 99, 108, 109, 113, 121, 122, 125, 129, 135
Easton Conservation Commission, 91
Easton Courier (newspaper), 108
Easton Garden Club, 87, 114, 115
Easton Historical Society, 86
Easton Homemakers, 102, 104
Easton: It's History (book), 17
Easton Library, 108
Easton Reservoir, 85
Easton's Open Space Task Force, 108
Efron, George, 52
Einstein, Albert, 49
Elliott, Dina, 106
Emmanuel Episcopal Church, 8
Environmental Protection Agency, 92, 137
Estock, Debra A., 108, 119
Everett Road, 68

Fairfield, 32, 87, 124
Fairfield-Citizen News 108, 119
Fairfield County, xi, 7, 40, 52, 54, 86, 90, 91, 110, 116, 120, 122, 124–126, 128, 129, 135, 138, 139

Fairfield Minuteman (newspaper), 108, 109
Falkenhagen, Princie, 87, 98, 99, 102, 106, 109, 110, 113, 122–124, 127, 128, 130, **132**
Fanton Hill Road, 60
Fanton, Sally, 17
Farnham, Thomas J., 15, 16
Federated Garden Clubs of Connecticut, 52
Ferragil Galleries, 52
Fitch, Frank, 42
Flirt Hill, 78
Ford Road, 61
Forum, 35
Four Corners, 86, 87, 132
Franklin, Bill, 68–78
Franklin, Olive, 68–78
Fredrik, Burry 20
Freedman, Sen. Judith, 126, 138
FXM Associates, 120

Garrity, 71
Gay, Duncan, 45, 50
Gay, Verne, 86, 106, 107
General Assembly, 116, 138
General Electric, 123
General Putnam Inn, 52
Gillies, Mrs. William, 64
Glocca Morra, 10
Godfrey, J. E., 44
Godfrey Road East, 13
Godfrey Street, 56, 67
Goldhorn, Mrs. L. H., 64
Goodhill Road, 25, 29
Gorham, Dr. Frank, 22–25, 139
Gorham, George M., 24
Gould, Stanley, 25
Gragan, Margaret P., 114, 115
Great Depression, 31, 32
Great Lakes, 18
Green, Nathaniel, 33
Greenwich, 8
Greenwich Art Gallery, 124
Griffin, (Jimmie) James, 42–44, 50, 54
Guidera, George, 93, 117, 124, 125, 128, 131–133, 136, 139
Gurney, A. R., 130

Haines, Joe, 81–83, 95
Hallock, Delia Hull, 13, 11
Hallock, Fred, 11, 25
Hallock, Henry, 23
Hallock, Jannette, 11
Harper, Mark, 100, 101
Harris, Charles S., 116
Harvey, Daniel, R., 44, 56, 59, 61
Hawley's Brook, 96, 97, 114, 117, 118, 120
Heath, Howard, 52
Heifetz, Patricia, 54
Hemlocks Reservoir, 83, 98
Henry Street Settlement, 47
Herald Magazine, 38–39, 43
Hodgins, Dick, 115, 119
Hoe, Arthur "Jim," 15, 17, 33–35
Hole in the Wall Gang Camp, 110
Holm, Cecil, 42
Horace C. Hurlbutt, Jr. School, 59
Hornstein, Harold, 104, 108, 109, 110
"How Things Were in Valley Forge," 10
Hoyt, Janey, 106
Hudson National Golf Club, 93
Hughes, Carolyn, 102
Hull, Bradley, 18
Hull Bradley, Clinton, 22, 23
Hull Cemetery, 80
Hull, Clinton, 13, 71, 139
Hull, Eva, 13
Hull, John, 17

Inman, Mary M., 46
International League for Peace and Freedom, 47

Johnson, Curtis, 6, 91, 98, 101, 102

Keedick, Lee, 42
Keeler, Samuel, 45
Keep Weston Rural, 139
Kelda, 7, 135, 137–139
Kent, 15
Kew Photo Labs, 12
Klein, Woody, 112
Kranyik, Louise, 90, 102
Kranyik, Robert, 7, 86, 87, 88, 101, 102, 104, 108–111, 118, 122, 123, 125, 129
Kupinse, William, 130

Index 147

Laurels, 54
Lee, Gypsy Rose, 54
Leff, David K., 126, 130
Legislative Office Building, 116
LePage, Bruce, 86, 93, 120, 127, 128, 129, 138
Litchfield County, 7, 16, 82, 138
Lomuscio, Christine, 112
Long Island Sound, 8, 32, 98, 117
Longshore Club House, 131
Love Letters (play), 130
Lucius Pond Ordway Devil's Den Preserve, 128
Lunch Box, 108
Lylburn, Margaret, 17, 18, 24
Lyons Plains Road, 8, 77, 125

Madden, Dick, 90, 94
Mark Twain Library, 64, 65
Massachusetts, 84, 120
McCollough, Peg, 66
McDonald, Tom, 95, 98
McGregor, Jack, 86
McKinney, Senator John, 126
McMansions, 8, 121
McMahon, Theresa, 44
Merrill, Mr., 71
Merritt Parkway, 45, 52
Metropolitan Opera, 131
Morehouse, Carrie and Charles, 44, 78
Morehouse, Charles Rowland, 77, 78
Morehouse, Frank, 44
Morehouse, Minerva, 76
Morton, Anson, 12, 70–71, 78–79, 83
Morton, Mrs., 71
Mother Teresa, 47

Naughton, James, 116
National Audubon Society, 116
National Child Labor Committee, 47
National Fairways, 5, 88, 93, 95, 107, 111, 125
National Trust for Public Lands, 123
Nature Conservancy, 115, 127, 128, 130, 137, 138
Neaton, Daniel, vi, 91
New Canaan, 8
New England Press Association, 108
New Haven, 16

Newman, Melissa, 109, 116
Newman, Nell, 109
Newman's Own, 110
Newman, Paul, vii, 6, 7, 109, 111, 112, 115, 117, 118, 122, 130, **131**, **132**, 137
Newtown Turnpike, 67, 76, 82
Nichols, Mrs. W. T., 64
New York–New Haven Railroad, 18
New York Times, 6, 89, 94, 102, 125
Noble, 106, 107
Norfield Road, 86, 125
Northern frogs, 100
Norton, Wesley, 44
Norwalk Hospital, 67
Norwalk-Newtown Turnpike, 56
Norwalk River, 54
Notre Dame Church, 104

Old Dimon Road, 56
Old Forge Road, 56
Open Space Task Force, 87
Oxford English Dictionary, 14

Partridge, Helen, 17
Patton, Dr. Steve, 128
Peck, Walter, 44, 45
Perry, Edgar, 42, 43, 50
Pida, Julia, 90
Pierson, Mrs., 75
Pinafore (light opera), 131
Pittsburgh, 18
Plymouth plantation, 14
Pop's Mountain, 8, 9, 54, 78, 79, 82
Pravato, Leonard, 12
Prince, William Meade, 42
Prosek, James, xi, 94, 96, 97, 106, 107, 122
Public Act 97–227, 116
puddling furnace, 14

Queen Elizabeth I, 14, 30

Redding, 1, 9, 39, 60, 64, 65, 82, 96, 100
Redding Glen, 45, 51, 64
Redford, Robert, 117–120
Revolutionary War, 16
Richardson, Dick and Judy, 106
Ricks, George, 80

Ricks, Sam, 80
Robinson, Grace, 67
Robotham, Mrs. Edward, 64
Roosevelt, Eleanor, 49
Roosevelt Elementary School, 66
Route 57, 11
Route 58, 100
Rowland, Charles, 25
Rowland, Elinor, 17
Rowland, Governor John G., 108, 118, 119, 127, 128, 138
Ruman, Irving, 69, 72

Salamon, Fannie, 24
Salvenius Fontinalis, 122
Samuel P. Senior Dam, 5, **6**, 8, **73**, **74**, **75**, **77**
Samuelson, Charles, 11
Samuelson, Gary, 11–13, 19, 20, 21, 22, 23, 25, 26, 29, 31
Samuelson, Janette, 13, 23
Samuel Staples auditorium, 129
Sanborn, James F., 50
Sanford, Aaron, 16
Sanford, Ephraim, 15
Sanford, Oliver, 15, 17
Sanford, Oliver C, 17
Sanford, Levi, 17
Sanford, Marion, 17
Sanford, William O., 17
Sanfordtown, 15
Saugatuck Reservoir, **vi**, 1, **2**, 9, **70**, 79, 81, 85, 89, 94, 97, 98, 100, 101, 138
Saugatuck River, xii, 8, 13, 15, 61, 114, 120
Saugatuck Saviours, 54
Saugatuck (River) Valley, 13, 14, 18, 31, 32, 34, 49, 52, **69**, 86, 131
Saugatuck Valley Association (SVA), **41**, 50, 52–55, **58**, 59, 60, 63, 65, 66, 104, 117
Saugatuck Valley Audubon Society, 116
Saugatuck Valley Defender (monthly newsletter), **53**
Save the Saugatuck, **48**, 132
Schlener, Denise, 127
Schmidt, Richard K., 113, 114
Senior, Samuel P., vii, xi, xii 34, 36, 37, 44, 55, 62, 63, 114

Shelton, 32
Sherman, 68
Sikorsky Aircraft, 32
Sokoloff, Mrs. Lyda, 64
Spanish (American) War, 16
Stafford Springs, 110
Stamford Water Company, 34
Staples High School, 64, 65
Steichen, Col. Edward, 61
Stowe Lumber Mill, 68
Strait, Don, 137
Stratford, 32
Stripp, John, Rep., 126
Sturges, John R., 17
Supreme Court, 125

Tallman, Arianne, 106, 120
Taylor, Rachel M., 46
Tennessee Valley Authority, 38
Tittle, Walter, 54
Todd, H. Stanley, 44
Toth Park, 120, 137
Trapp Falls, 50
Treadwell, Walter, 17
Trout: An Illustrated History (book), 94, 96
Trout Brook (Valley), xi; 5, 6, 7, 9, 54, 81, 85, 86, 88, 89, 91–94, **95**, **96**, **97**, 98, 100, **103**, 104–107, 110, 112, 113, 115–117, 120, 121–128, **129**, 130, 131, 135, 136, 139
Trout Brook Valley Preservation Update (twice annual newsletter), **105**
tunnel construction, **78**

U.S. Geological Survey, 61

Valley Forge, xi, **xvi**, 3, 5, 7, 14, 15, 20, 24, 27, 29, 31, 36–38, 39, 43, 44, 46, 55, 60, 63, 68, 83, 99, 100, 111, 115–117, 128, 131, 132, 137; dam, **36**; farm, **21**; fishers, **30**; house circa 1830, **25**; hunters, **26**, **29**; sawmill, **38**; school children, **24**; school district, 66
Valley Forge Glen, **18**
Valley Forge Road, 8, 9, 56, 76, 83, 97, 124
Valley Forge School, 20, **23**, 23–25, 66

Van Riper, George, 45
Visiting Nurse Society, 47

Wald, Lillian, 47, 49–51, 63, 110
War of 1812, 16
Waterbury, George, 56
Weston, 1, 6, 8, 9, 15, 39, 41, 48, 61, 64, 83, 86, 87, 93, 97, 101, 117, 131, 135
Weston Boy Scouts, 9
Weston Commission for the Arts, 10
Weston's Emergency Services, 10
Weston Historical Society, Inc., 2, 3, 6, 10, 16, 18, 19, 20, 23, 33, 34, 36, 38, 41, 48, 53, 58, 69, 70, 73–75, 77, 78, 132, 139
Weston League of Women Voters, 135
Weston Road, 125
Weston: The Forging of A Connecticut Town (book), 15
Weston Town Hall, 12, 17, 131
Weston Town Meeting, 40
Westport, 61, 64, 87, 125
Westport Chamber of Commerce, 51
Westport Country Playhouse, 130
Westport-Danbury Road, 56
Westporter-Herald (newspaper), 16, 17, 36, 37, 40, 43, 44, 45, 49–56, 60–67

Westport Garden Club, 52
Westport Magazine, 135, 137
Westport Minuteman (newspaper), 108
Westport News, 104, 108, 109, 112, 113, 115, 117, 119, 135
Westport Rotary, 36
Westport YMCA, 36
Wheeler Foundry, **33**, 35
Wheeler, Henry B., 18
Whitlock, H. D. Adlebert, 10, 12, 19, 20, 21, 22, 25–30, 66, 100
Wilder, Frank L., 45
Williams, Eva Jane, 23
Wister II, Mrs. William, 64
Women's Trade Union League, 47
Women's Peace party, 47
Woodward, Joanne, 130
Woods, Tiger, 102
Woolson, Lucie, 46
World War I, 32, 47, 66
World War II, 66
Wright, George, 52
Wynne, Judge Kenneth, 43, 45, 50

Yale University, 101
Young, John Orr, 36, 56, 59, 62, 63
Young and Rubicam, 36